Lecture Notes in Bioinformatics　　　　11488

Subseries of Lecture Notes in Computer Science

More information about this series at http://www.springer.com/series/5381

Ian Holmes · Carlos Martín-Vide ·
Miguel A. Vega-Rodríguez (Eds.)

Algorithms for Computational Biology

6th International Conference, AlCoB 2019
Berkeley, CA, USA, May 28–30, 2019
Proceedings

 Springer

Editors
Ian Holmes ⓘ
University of California, Berkeley
Berkeley, CA, USA

Carlos Martín-Vide ⓘ
Rovira i Virgili University
Tarragona, Spain

Miguel A. Vega-Rodríguez ⓘ
University of Extremadura
Cáceres, Spain

ISSN 0302-9743 ISSN 1611-3349 (electronic)
Lecture Notes in Bioinformatics
ISBN 978-3-030-18173-4 ISBN 978-3-030-18174-1 (eBook)
https://doi.org/10.1007/978-3-030-18174-1

LNCS Sublibrary: SL8 – Bioinformatics

This Springer imprint is published by the registered company Springer Nature Switzerland AG
The registered company address is: Gewerbestrasse 11, 6330 Cham, Switzerland

Preface

These proceedings contain the papers that were presented at the 6th International Conference on Algorithms for Computational Biology (AlCoB 2019), held in Berkeley, California, USA, during May 28–30, 2019.

The scope of AlCoB includes topics of either theoretical or applied interest, namely:

- Sequence analysis
- Sequence alignment
- Sequence assembly
- Genome rearrangement
- Regulatory motif finding
- Phylogeny reconstruction
- Phylogeny comparison
- Structure prediction
- Compressive genomics
- Proteomics: molecular pathways, interaction networks, mass spectrometry analysis
- Transcriptomics: splicing variants, isoform inference and quantification, differential analysis
- Next-generation sequencing: population genomics, metagenomics, metatranscriptomics, epigenomics
- Genome CD architecture
- Microbiome analysis
- Cancer computational biology
- Systems biology

AlCoB 2019 received 30 submissions. Most papers were reviewed by three Program Committee members. There were also a few external reviewers consulted. After a thorough and vivid discussion phase, the committee decided to accept 15 papers (which represents an acceptance rate of about 50%). The conference program included five invited talks and some poster presentations of work in progress.

The excellent facilities provided by the EasyChair conference management system allowed us to deal with the submissions successfully and handle the preparation of these proceedings in time.

We would like to thank all invited speakers and authors for their contributions, the Program Committee and the external reviewers for their cooperation, and Springer for its very professional publishing work.

March 2019
Ian Holmes
Carlos Martín-Vide
Miguel A. Vega-Rodríguez

Organization

AlCoB 2019 was organized by the University of California, Berkeley, USA, and the Institute for Research Development, Training and Advice (IRDTA), Brussels/London, Belgium/UK.

Program Committee

William Pearson	University of Virginia, USA
Matteo Pellegrini	University of California, Los Angeles, USA
Mihaela Pertea	Johns Hopkins University, USA
Steve Rozen	Duke-NUS Medical School, Singapore
David Sankoff	University of Ottawa, Canada
Russell Schwartz	Carnegie Mellon University, USA
Wing-Kin Sung	National University of Singapore, Singapore
Alfonso Valencia	Barcelona Supercomputing Centre, Spain
Arndt von Haeseler	Center for Integrative Bioinformatics Vienna, Austria
Kai Wang	Children's Hospital of Philadelphia, USA

Additional Reviewers

Wout Bittremieux	Shaoheng Liang
Viraj Deshpande	Vakul Mohanty
Petko Fiziev	Jacob Schreiber
Markus Fleischauer	Yifei Shen
Songling Li	Ritambhara Singh

Organizing Committee

Ian Holmes (Co-chair)	University of California, Berkeley, USA
Sara Morales	IRDTA, Brussels, Belgium
Manuel Parra-Royón	University of Granada, Granada, Spain
David Silva (Co-chair)	IRDTA, London, UK
Miguel A. Vega-Rodríguez	University of Extremadura, Cáceres, Spain

Abstracts of Invited Talks

Exploring Phenotypic Heterogeneity Across Tissues and Conditions with Network-Based Approaches

Teresa M. Przytycka

National Center of Biotechnology Information, National Library of Medicine,
NIH, Bethesda, MD 20894, USA

Phenotypic heterogeneity is assumed arise as the result of a combination of genetic, epigenetic, and environmental factors and the stochastic nature of biochemical processes, such as gene expression events, during the development. In the last few years, my group has studied phenotypic heterogeneity in two different contexts: in the context of cancer [2–5, 7, 8] and in the context of the model organism – *Drosophila melanogaster* [10, 11].

Functional interaction networks, that is networks whose edges represent functional relationships between genes, provide an important context for studies of organismal phenotypes. A network-centric view of genotype-phenotype relation proposes that perturbing functionally related genes is likely to lead to similar phenotypes. Indeed, network-centric approaches have proven to be helpful for finding genotypic causes of diseases, classifying disease into subtypes, and identifying drug targets [1, 5]. To support such pathway-centric perspective, algorithms that leverage biological networks to advance the understanding of phenotypic heterogeneity are necessary.

In our earlier study, building on the set cover approach, we have developed network-based approaches to identify pathways dysregualted in cancer [4, 7]. In contrast, focusing on uncovering pathways mediating the relation between somatic mutations and dysregulatcd gene expression modules, we utilised a network flow approach [6, 8]. Complementing these studies, our recently developed method, BeWith, utilises Integer Linear Programming and an integrated analysis of mutual exclusivity, co-occurrence and functional interaction networks to uncover the relationships between mutated gene modules. As expected such mutated gene modules often underline specific cancer sub-types [3].

In addition to studies on dysregulated pathways in cancer and cancer subtypes, network based approaches can also shed light on other phenotypes such as drug response. Towards this end we have recently developed NETPHLIX - an algorithm to identify mutated subnetworks that are associated with a continuous phenotype. Subsequently, we utilised NETPHLIX to identify mutated gene networks that are associated with response to drugs. Another recently emerged phenotype in cancer studies is the presence and strength of the so-called mutational signatures. Mutational signatures are indicative of mutagenic processes that have been active in the given patient. Such processes are often triggered by genetic causes such as a dysfunctional DNA repair pathway. Understanding the mechanism behind the emergence of a particular mutational signature is challenging since increased mutagenic activities leads to an increased amount of passenger mutations making it difficult to untangle the cause from the effect. Using NETPHLIX, we were able to identify mutated sub-networks associated with several mutational signatures in breast cancer [9].

In contrast to functional interaction networks, gene regulatory networks (GRN) summarise regulatory relationships between transcription factors (TF) and the gens that they regulate. GRN regulate maintenance of cell type specific states, response to stress, and other cell functions. Thus phenotypic differences can be potentially explained by differences in gene regulation. However methods to infer GRN are typically context-agnostic. To address this challenge, we have recently introduced a novel computation method NetREX that given a context-agnostic network as a prior and context specific expression data (for example data for a healthy and a disease tissue), constructs context-specific GRNs by rewiring the prior network [11]. Comparative analysis of such networks can provide yet another window to study phenotypic differences.

We conclude that network based approaches, supported by a variety of algorithmic approaches can provide important stepping stone towards understating phenotypic heterogeneity.

Acknowledgements. I would like to acknowledge all the collaborators of the work discussed in this talk. Particular thanks to the current and former members of my group: Yoo-Ah Kim, Yijie Wang, Damian Wojtowicz, Phoung Dao, Jan Hoinka, Dang-Yon Cho, Raheleh Salari our visiting group member Rebecca Sarto-Basso, and our many collaborators including Roded Sharan, Fabio Vandin, Brian Oliver, Dorit S Hochbaum, Stefan Wuchty, Hang Noh Lee, Max Leiserson and all the other collaborators listed in the bibliography of this talk. The research in Przytycka's group is supported by the Intramural Research Programs of the National Library of Medicine at National Institutes of Health, USA.

References

1. Cho, D.Y., Kim, Y.A., Przytycka, T.: Network biology approach to complex diseases. PLoS Comput. Biol. **8**(12) (2012). https://doi.org/10.1371/journal.pcbi.1002820

2. Cho, D.Y., Przytycka, T.M.: Dissecting cancer heterogeneity with a probabilistic genotype-phenotype model. In: Deng, M., Jiang, R., Sun, F., Zhang, X. (eds.) RECOMB 2013. LNCS, vol. 7821, pp. 30–31. Springer, Heidelberg (2013). https://doi.org/10.1007/978-3-642-37195-0_3

3. Dao, P., Kim, Y.A., Wojtowicz, D., Madan, S., Sharan, R., Przytycka, T.: BeWith: a between-within method to discover relationships between cancer modules via integrated analysis of mutual exclusivity, co-occurrence and functional interactions. PLoS Comput. Biol. **13**(10) (2017). https://doi.org/10.1371/journal.pcbi.1005695

4. Kim, Y.A., Cho, D.Y., Dao, P., Przytycka, T.: MEMCover: integrated analysis of mutual exclusivity and functional network reveals dysregulated pathways across multiple cancer types. Bioinformatics **31**(12) (2015). https://doi.org/10.1093/bioinformatics/btv247

5. Kim, Y.A., Cho, D.Y., Przytycka, T.: Understanding genotype-phenotype effects in cancer via network approaches. PLoS Comput. Biol. **12**(3) (2016). https://doi.org/10.1371/journal.pcbi.1004747

6. Kim, Y.A., Przytycki, J., Wuchty, S., Przytycka, T.: Modeling information flow in biological networks. Phys. Biol. **8**(3) (2011). https://doi.org/10.1088/1478-3975/8/3/035012
7. Kim, Y.A., Salari, R., Wuchty, S., Przytycka, T.: Module cover - a new approach to genotype-phenotype studies. In: 18th Pacific Symposium on Biocomputing, PSB 2013 (2013)
8. Kim, Y.A., Wuchty, S., Przytycka, T.: Identifying causal genes and dysregulated pathways in complex diseases. PLoS Comput. Biol. **7**(3) (2011). https://doi.org/10.1371/journal.pcbi.1001095
9. Kim, Y.A., et al.: Network-based approaches elucidate differences within APOBEC and clock-like signatures in breast cancer. bioRxiv, p. 568568, March 2019. https://doi.org/10.1101/568568, https://www.biorxiv.org/content/10.1101/568568v1?rss=1
10. Lee, H., et al.: Dosage-dependent expression variation suppressed on the Drosophila male X chromosome. G3: Genes Genomes Genet. **8**(2) (2018). https://doi.org/10.1534/g3.117.300400
11. Wang, Y., Cho, D.Y., Lee, H., Fear, J., Oliver, B., Przytycka, T.: Reprogramming of regulatory network using expression uncovers sex-specific gene regulation in Drosophila. Nat. Commun. **9**(1) (2018). https://doi.org/10.1038/s41467-018-06382-z

New Divide-and-Conquer Techniques
for Large-Scale Phylogenetic Estimation[1]

Tandy Warnow

Department of Computer Science, University of Illinois at
Urbana-Champaign, 201 N. Goodwin Ave, Urbana, IL, 61801, USA
warnow@illinois.edu

Over the last years, the availability of genomic sequence data from thousands of different species has led to hopes that a phylogenetic tree of all life might be achievable. Yet, the most accurate methods for estimating phylogenies are heuristics for NP-hard optimization problems, many of which are too computationally intensive to use on large datasets. Divide-and-conquer approaches have been proposed to address scalability to large datasets that divide the species into subsets, construct trees on subsets, and then merge the trees together. Prior approaches have divided species sets into overlapping subsets and used supertree methods to merge the subset trees, but limitations in supertree methods suggest this kind of divide-and-conquer approach is unlikely to provide scalability to ultra-large datasets. Recently, a new approach has been developed that divides the species dataset into disjoint subsets, computes trees on subsets, and then combines the subset trees using auxiliary information (e.g., a distance matrix). Here, we describe these strategies and their theoretical properties, present open problems, and discuss opportunities for impact in large-scale phylogenetic estimation using these and similar approaches.

Acknowledgements. This work was supported in part by NSF grant CCF-1535977. I also wish to thank Erin Molloy and Thien Le for helpful comments on the manuscript.

---

[1] Supported by the University of Illinois at Urbana-Champaign.

Contents

Invited Talk

New Divide-and-Conquer Techniques for Large-Scale
Phylogenetic Estimation.................................... 3
   Tandy Warnow

Biological Networks and Graph Algorithms

New Polynomial-Time Algorithm Around the Scaffolding Problem........ 25
   Tom Davot, Annie Chateau, Rodolphe Giroudeau, and Mathias Weller

Enumerating Dominant Pathways in Biological Networks by Information
Flow Analysis 39
   Ozan Kahramanoğulları

Comparing Different Graphlet Measures for Evaluating Network Model
Fits to BioGRID PPI Networks.............................. 52
   Sridevi Maharaj, Zarin Ohiba, and Wayne Hayes

Graph-Theoretic Partitioning of RNAs and Classification of Pseudoknots.... 68
   Louis Petingi and Tamar Schlick

PathRacer: Racing Profile HMM Paths on Assembly Graph 80
   Alexander Shlemov and Anton Korobeynikov

Genome Rearrangement, Assembly and Classification

A Uniform Theory of Adequate Subgraphs for the Genome Median,
Halving, and Aliquoting Problems............................ 97
   Pavel Avdeyev, Maria Atamanova, and Max A. Alekseyev

Lightweight Metagenomic Classification via eBWT.................. 112
   Veronica Guerrini and Giovanna Rosone

MULKSG: *MUL*tiple *K* *S*imultaneous Graph Assembly 125
   Christopher Wright, Sriram Krishnamoorty, and Milind Kulkarni

Counting Sorting Scenarios and Intermediate Genomes
for the Rank Distance 137
   João Paulo Pereira Zanetti, Leonid Chindelevitch, and João Meidanis

Generalizations of the Genomic Rank Distance to Indels   152
    João Paulo Pereira Zanetti, Leonid Chindelevitch, and João Meidanis

Sequence Analysis, Phylogenetics and Other Biological Processes

Using INC Within Divide-and-Conquer Phylogeny Estimation   167
    Thien Le, Aaron Sy, Erin K. Molloy, Qiuyi (Richard) Zhang, Satish Rao,
    and Tandy Warnow

Predicting Methylation from Sequence and Gene Expression Using Deep
Learning with Attention.   179
    Alona Levy-Jurgenson, Xavier Tekpli, Vessela N. Kristensen,
    and Zohar Yakhini

A Mathematical Model for Enhancer Activation Kinetics During
Cell Differentiation   191
    Kari Nousiainen, Jukka Intosalmi, and Harri Lähdesmäki

Transcript Abundance Estimation and the Laminar Packing Problem   203
    Atif Rahman and Lior Pachter

Efficient Algorithms for Finding Edit-Distance Based Motifs   212
    Peng Xiao, Xingyu Cai, and Sanguthevar Rajasekaran

Author Index   225

Invited Talk

New Divide-and-Conquer Techniques
for Large-Scale Phylogenetic Estimation

Tandy Warnow(✉) (iD)

Department of Computer Science, University of Illinois at Urbana-Champaign,
201 N. Goodwin Ave, Urbana, IL 61801, USA
warnow@illinois.edu

Abstract. Over the last years, the availability of genomic sequence data
from thousands of different species has led to hopes that a phylogenetic
tree of all life might be achievable. Yet, the most accurate methods for
estimating phylogenies are heuristics for NP-hard optimization problems,
many of which are too computationally intensive to use on large datasets.
Divide-and-conquer approaches have been proposed to address scalability
to large datasets that divide the species into subsets, construct trees
on subsets, and then merge the trees together. Prior approaches have
divided species sets into overlapping subsets and used supertree methods
to merge the subset trees, but limitations in supertree methods suggest
this kind of divide-and-conquer approach is unlikely to provide scalability
to ultra-large datasets. Recently, a new approach has been developed
that divides the species dataset into disjoint subsets, computes trees on
subsets, and then combines the subset trees using auxiliary information
(e.g., a distance matrix). Here, we describe these strategies and their
theoretical properties, present open problems, and discuss opportunities
for impact in large-scale phylogenetic estimation using these and similar
approaches.

Keywords: Inferring the evolutionary phylogeny of species ·
Incomplete lineage sorting · Gene trees · Species trees ·
Divide-and-conquer · Absolute fast converging methods ·
Statistical consistency

1 Introduction

Phylogenies are graphical models for how a set S of species, genes, or other
"taxa" evolved from a common ancestor. In its most common. usage, a phylogeny
is assumed to be a rooted binary tree, with leaves labelled by the taxa in the
set S, and with internal nodes representing ancestral taxa. Phylogenetic trees
are used to provide insight into many biological research questions, including
how genes co-evolve, how organisms adapt to their environments, how humans

---

Supported by the University of Illinois at Urbana-Champaign.

I. Holmes et al. (Eds.): AlCoB 2019, LNBI 11488, pp. 3–21, 2019.
https://doi.org/10.1007/978-3-030-18174-1_1

migrated across the globe, etc. Because of these and other uses, the inference of phylogenetic trees is a major step in many biological studies.

The inference of the phylogeny given the input set is generally approached as a statistical estimation problem, where the objective is often to find the model tree that is most likely to have generated the observed data. Since the maximum likelihood problems are generally NP-hard, heuristics are used to search for good solutions. However, standard heuristic search strategies do not scale well to large datasets, with the result that phylogeny estimation on large datasets is very computationally intensive.

Divide-and-conquer strategies have been developed to improve the scalability of phylogeny estimation methods that have the following approach: the dataset is divided into overlapping subsets, trees are constructed on the subsets, and then the trees are combined into a tree on the full dataset using a supertree method [34]. Despite the promising performance of these divide-and-conquer strategies, limitations in currently available supertree methods suggest that these strategies will not provide good scalability to the large and utlra-large datasets that are of interest in biological research.

In this paper we describe a new type of divide-and-conquer strategy for use with phylogeny estimation methods. Unlike prior strategies, the dataset is divided into disjoint subsets, trees are computed on these subsets, and then the trees are combined into a tree on the full set of species. Since the trees are disjoint, the merger of these disjoint trees requires auxiliary information, such as a distance matrix relating the species to each other. Thus, the disjoint tree merger (DTM) problem is of interest in its own right.

To date, only two methods have been developed with this strategy, each based on a different DTM method. In this paper, we describe two DTM strategies, NJMerge [31] and INC [60], and their use within divide-and-conquer strategies. The results so far are promising, but much still needs to be done.

The rest of the paper is organized as follows. We begin with background about phylogeny estimation in Sect. 2, including two different types of phylogeny estimation problems, gene tree estimation and species tree estimation. We present DTM methods and their theoretical results in Sect. 3. Results using DTM methods within divide-and-conquer pipelines are discussed in Sect. 4. Finally, we conclude with a discussion about future work and open problems in Sect. 5.

2   Phylogeny Estimation

2.1   Gene Tree Estimation

Phylogeny estimation is most typically performed on DNA sequences for a single gene (or, more generally, a single locus) and assumes that the sequences evolved down a common "model tree" under a stochastic model of sequence evolution that includes substitutions of nucleotides by other nucleotides as well as insertions and deletions (i.e., "indels") of nucleotides. A simple example of such a model is the Juke-Cantor (JC69, [15]) model for single site evolution, in which the nucleotide at the root is selected uniformly at random and then evolves down the tree. Each edge e in the tree has a substitution probability $p(e)$ with $0 < p(e) < 0.75$, which specifies the probability that the site will change state on the edge. If a site changes on edge e, then it selects its new state from the remaining three nucleotides with equal probability. A JC69 model tree is thus defined by the pair (T, Θ), where T represents the rooted binary tree with leaves labelled by the taxa in S and Θ representing the numeric parameters on the tree (i.e., the probabilities $p(e)$ of change on the edges in T). More complex models, such as the Generalised Time Reversible (GTR) model [48], have also been developed, to better model biological sequence evolution. The GTR model contains the JC69 model and is the most complex of the site evolution models used in practice; like the JC69 model, it does not include indels, and instead assumes all evolution is restricted to substitutions. Since biological datasets evolve with indels, a phylogenetic analysis under these models generally operates in two steps: (1) the sequences are first aligned and (2) the computed multiple sequence alignment is analyzed using the assumed model, treating the gaps as missing data.

The challenge is then to infer the true (or model) tree from the sequences that it produced. For example, a maximum likelihood (ML) approach can be taken, which seeks the model tree (i.e., rooted binary tree with numeric parameters of evolution that specify the stochastic process) that has the highest probability of generating the observed data [11]. Because ML is NP-hard [38], heuristics (typically based on local search strategies, such as NNI (nearest neighbor interchanges) moves, are used to search for good solutions within tree space. However, local search strategies are not that effective at finding good solutions to NP-hard optimization problems, and the number of different binary rooted trees grows exponentially with the number of leaves. Hence, it is not surprising that the current leading ML heuristics, such as RAxML [44] and IQTree [35], are computationally intensive on large datasets.

One alternative to heuristic search strategies are distance-based methods, which operate as follows. First, distances are computed between the taxa using the properties about the assumed model [46], and then trees can be computed on these distances. For the standard models of sequence evolution, statistical methods for calculating the matrix of pairwise distances have been developed that are guaranteed to converge to *additive matrices* for the true tree, where D is said to be additive on T if there is a way of assigning non-negative weights to the edges of T so that D_{ij} is the sum of the edge weights on the path in T between taxa i and j. We provide this definition formally, so we can refer to it later:

Definition 1. *Let T be a binary tree on leafset $1, 2, \ldots, n$ and with positive edge weights defined by $w : E(T) \rightarrow \mathbb{R}^+$. Then we define a matrix A of leaf-to-leaf distances by setting $A[i, j]$ to be the sum of the weights of the edges on the path between leaves i and j. The matrix A is then said to be additive on T. More generally, any matrix B that can be realized as path distances in T for some positive edge-weighting of T is said to be* **additive on** T.

Distance-based approaches are typically polynomial time, since the calculation of the matrix of pairwise distances requires only polynomial time and the calculation of a tree given the matrix also requires only polynomial time. The best known method for calculating trees from distance matrices is neighbor joining (NJ, [42]), which uses $O(n^3)$ time; however, other methods also exist, including $O(n^2)$ algorithms within the FastME [21] package. These distance-based methods are thus inherently faster than the heuristics for maximum likelihood.

When treating phylogeny construction as a statistical estimation problem, a basic question is whether a given method is statistically consistent under the model, which means that as the amount of data increases the tree returned by the data will converge to the true tree (in an unrooted form). Since the models we consider are time-reversible, we will attempt to infer the unrooted version of T, which we refer to as T^u. As an example, for the JC69 model, a method M would be said to be statistically consistent if, for all JC69 model trees (T, Θ) and $\epsilon > 0$, there is a sequence length K such that given set S of sequences of length at least K the probability that $M(S) = T^u$ is at least $1 - \epsilon$.

Many methods are known to be statistically consistent under the GTR model, and hence under its submodels; for example, neighbor joining and many other distance-based methods are statistically consistent, as is maximum likelihood (and also Bayesian methods); see [1] for the proof for neighbor joining and [12,56,58] for more about these issues. On the other hand, distance-based tree estimation is generally not as accurate as maximum likelihood tree estimation on simulated datasets [54], suggesting that computationally intensive heuristics in ML codes may be necessary.

2.2 Species Tree Estimation

The discussion so far has assumed that all the sites evolve down a single model tree. However, genomes evolve with processes, such as incomplete lineage sorting and gene duplication and loss, that cause different parts of the genome to evolve under different trees [26]. As a result, although the inference of the phylogenetic tree for a single genomic region can be reasonably addressed under models that assume that all the sites evolve down a single model tree, the same assumption is not generally valid when working with genome-scale evolution. More generally, multi-locus phylogeny estimation (which means the estimation of an evolutionary tree using sites from different loci) requires different models of evolution and different methods for estimating under these models. In other words, there is a distinction between "gene trees", which represent the evolution of a single gene (or genomic region, also called a locus) and a "species tree", which represents the evolution of the species as reflected in their genomes.

Species tree estimation is thus more complicated than gene tree estimation. One of the active research areas is the development of methods that can infer species trees from multi-locus datasets fwhen gene trees can differ from the species tree, and from each other, due to incomplete lineage sorting (ILS). In this context, gene trees are seen as evolving within a species tree under the multi-species coalescent (MSC) model [16], with "failure to coalesce" on a branch (resulting from rapid speciation and/or very large effective population sizes) making it possible for gene trees to differ from the species tree. Thus, sequence evolution can be described by gene trees evolving within a species tree T under the MSC model and sequences evolve down each gene tree under the GTR model. Under this hierarchical MSC+GTR model, the task is to estimate T^u given the multi-locus sequence dataset.

A phylogeny estimation method is said to be statistically consistent under the MSC+GTR model if, for all MSC+GTR model species trees (T, θ) (where θ provides the numeric parameters) as the number of sites per locus and the number of loci both increase, the probability of recovering T^u converges to 1. Methods that are statistically consistent under the MSC+GTR model have been developed, including some that are polynomial time and reasonably computationally efficient in practice (e.g., ASTRAL [28,29,59], ASTRID [51], and NJst [23]), others that are polynomial time but more computationally intensive (e.g., SVDquartets [7,8], SVDquest [53]), and others that use Bayesian MCMC and so cannot be used on large numbers of species due to slow convergence rates (e.g., *BEAST [13], see [62]). In contrast, standard maximum likelihood analysis, which concatenates all the alignments into a single "super-alignment", treats all the sites as evolving down a single tree, and then seeks the model tree that maximizes the probability of generating the observed data, is *not* statistically consistent under the MSC+GTR model [39,40].

Furthermore, these "concatenation" analyses can have high error on simulated datasets [18,32]. Nevertheless, even the best performing statistically consistent methods, such as ASTRAL, can be less accurate than concatenation analyses under some conditions [32]. Therefore, in practice evolutionary biologists use several types of methods (typically a concatenation analysis, often based on RAxML, and ASTRAL or some other similarly fast coalescent-based method) to estimate species trees.

RAxML and other ML heuristics are unable to find global optima in most cases, and instead find local optima that may or may not be particularly good solutions, and the time to converge to a local optimum can be very large; in addition, ML heuristics often need a lot of memory (but this is a function of the input). Fortunately, many of these heuristics record intermediate solutions, and so can be stopped after some elapsed time and the best tree returned. ASTRAL and other similar methods run in polynomial time, but even so can fail to return any tree at all if their analysis does not complete (i.e., they do not perform a heuristic search, and so do not produce any intermediate trees). In other words, all of these methods can be challenging to run on large datasets.

2.3 Divide-and-Conquer Strategies

Divide-and-conquer is a standard algorithmic technique that has been used in phylogeny estimation. For example, the "disk-covering methods" (see [14,19,34, 55,56]) divide a set of taxa, each given by a DNA sequence, into subsets that are overlapping, then constructs trees on the subsets using a preferred "base method", and then combine the subset trees together using a supertree method.

Supertree methods [36,41,56] take as input a set of unrooted trees, each leaf-labelled by a subset of the species set S and returns an unrooted phylogenetic tree T on the full set S. (Note that we assume here, as elsewhere in the paper, that each tree has at most one leaf for each species in any input tree.) Supertree construction can be addressed using optimization criteria (e.g., find a tree T that minimizes the total Robinson-Foulds distance to the input trees [2,6,52]), but since tree compatibility is NP-complete [45], all optimization problems are NP-hard. This type of divide-and-conquer approach can produce high accuracy under some conditions [3,34], but due to limitations in current supertree methods [57], these divide-and-conquer approaches may not be helpful for scaling species tree estimation to ultra-large datasets.

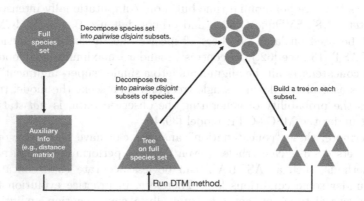

Fig. 1. Generic form of divide-and-conquer pipelines using Disjoint Tree Merger (DTM) methods. The input is a set of species along with some auxiliary information (e.g., a distance matrix, a multiple sequence alignment, etc.). In the first iteration, the species set is divided into pairwise-disjoint subsets, trees are constructed on the subsets, and then merged together using the selected DTM method and the auxiliary information. All iterations after the first divide the species set into subsets using the tree produced by the previous iteration, and otherwise use the same approach. The process ends when a stopping rule is triggered, which can be the selected number of iterations, obtaining a good score to some optimality criterion, etc. Note that the subsets that are produced are pairwise-disjoint, which is why DTM methods must be used rather than supertree methods. However, this approach is closely related to DACTAL [34], which had the same overall format but produced overlapping subsets, and then merged the trees together using supertree methods.

Figure 1 presents a new type of divide-and-conquer approach for phylogeny estimation that gets around the challenges inherent in using supertree methods: rather than dividing into overlapping subsets and then combining trees on the subsets using supertree methods, the species set is divided into disjoint subsets and then trees on the subsets are merged together using the input sequence data for the species. If desired, this kind of approach can iterate, where each iteration begins with the tree from the previous iteration, divides the species set into disjoint subsets using that tree, constructs trees on the subsets, and then merges them together into a tree on the full dataset.

Thus, a key part of this approach is the step where a set of leaf-disjoint trees are merged into a single tree on the full set of species, i.e., the Disjoint Tree Merger (DTM) problem, defined below. NJMerge [30,31] and INC [60] are the two methods that have been developed for the Disjoint Tree Merger Problem.

3 Disjoint Tree Merger

3.1 The DTM Problem

We will assume throughout this paper that every tree (other than model trees) is an unrooted binary tree with leaves labelled bijectively by subsets of the species set S (thus, each species labels at most one leaf in any given tree). We will also assume that $\mathcal{T}$ is a set of trees. D will denote a dissimilarity matrix on S, which means that D is symmetric and zero on the diagonal, but we do not require that D satisfies the triangle inequality (because distances between species in phylogenetics typically do not satisfy the triangle inequality). We let $\mathcal{L}(t)$ denote the leafset of tree t and we let t_X denote the homeomorphic subtree of t induced by X (so that nodes of degree two are suppressed).

Definition 2. *Let $\mathcal{T}$ be a set of unrooted binary trees, each on a subset of S and let $\cup_{t \in \mathcal{T}} \mathcal{L}(t) = S$. Then tree T with leafset S is said to be a* **compatibility supertree** *for $\mathcal{T}$ if $T|_{\mathcal{L}(t)} = t$ for all $t \in \mathcal{T}$.*

The **Disjoint Tree Merger (DTM) Problem** is defined as follows.

- **Input:** A pair $(\mathcal{T}, D)$, where $\mathcal{T}$ is a set of leaf-disjoint unrooted binary trees, $\cup_{t \in \mathcal{T}} \mathcal{L}(t) = S$, and D is a dissimilarity matrix on S
- **Output:** An unrooted binary tree T with leafset S that is a compatibility supertree for $\mathcal{T}$ (see Definition 2).

Because the trees in $\mathcal{T}$ are leaf-disjoint, a compatibility supertree always exists: create a node v and make it adjacent to some node in every tree in $\mathcal{T}$. Note that the trees in $\mathcal{T}$ define convex subtrees within the compatibility supertree (i.e., given any two leaves x, y in a tree in $\mathcal{T}$, the path between x, y in the compatibility supertree does not pass through any vertex in any other tree in $\mathcal{T}$). However, for every input $\mathcal{T}$, other compatibility supertrees exist that do not have this convexity property. For example, consider the caterpillar tree $(1, 2, 3, 4, \ldots, 2n)$ constructed by taking the path $v_2, v_3, \ldots, v_{2n-1}$ and making each node on the path adjacent to one or two leaves, as follows: v_i is adjacent to leaf i (for $i = 3$ to $2n - 2$), v_2 is adjacent to leaves 1 and 2, and v_{2n-1} is adjacent to leaves $2n - 1$ and $2n$. This caterpillar tree is a compatibility supertree for caterpillar trees $(1, 3, 5, 7, 9, \ldots, 2n - 1)$ and $(2, 4, 6, 8, \ldots, 2n)$. Note that this compatibility supertree does not have the convexity property, and instead *blends* the trees in $\mathcal{T}$ together. Thus, we can distinguish between *blended* compatibility supertrees and *unblended* compatibility supertrees.

Recall that our objective is to find a compatibility supertree that is close to the true tree. Since the true tree is unknown, we can address this objective using simulation studies or through theory, but in each case we will assume that the trees in $\mathcal{T}$ are species trees that we estimate using multi-locus sequence data. However, the best empirical (and theoretical) accuracy requires the use of blended compatibility supertrees.

3.2   NJMerge

NJMerge [30] is an algorithm for the DTM problem that is a modification of the NJ algorithm from [42]. Thus, the input to NJMerge is a pair $(\mathcal{T}, D)$, where $\mathcal{T}$ is a set of leaf-disjoint constraint trees and D is an $n \times n$ dissimilarity matrix. Here we briefly summarize the NJ algorithm, and then describe how NJMerge modifies the algorithm to address the constraints in $\mathcal{T}$.

NJ takes as input an $n \times n$ dissimilarity matrix D, and computes a second $n \times n$ matrix Q using D (see [1,42,56] for details). It then finds the pair i, j that minimizes $Q[i, j]$, and makes i and j siblings. It replaces the pair i, j in D by a new taxon x and calculates the distance from x to every other taxon $k \neq i, j$; as a result, D has now $n - 1$ rows and columns. Note that as the algorithm progresses, each taxon represents a rooted binary tree on some of the original set of leaves. The algorithm repeats until D has only two rows and columns, at which point the two remaining taxa are made into a sibling pair, and the algorithm returns the rooted tree on the original set of leaves. This rooted tree is generally interpreted as an unrooted tree on the original set of leaves.

NJMerge follows the same approach, but with modifications so as to avoid accepting siblinghood proposals that violate the current set of constraint trees or might make the set of modified constraint trees incompatible. That is, the consequence of making i and j siblings may be neutral (for example, i and j are siblings in one of the constraint trees, and neither appears in any other constraint tree) or it may potentially lead to incompatibility. For example, suppose that a few iterations of NJMerge, the set $\mathcal{T}$ contains trees $12|34, 23|45, 34|56$; note that $\mathcal{T}$ is compatible and that the caterpillar tree $(1, 2, 3, 4, 5, 6)$ is the unique compatibility supertree for $\mathcal{T}$. Now suppose that $1, 6$ is the pair that minimizes $Q[i, j]$. Making 1 and 6 siblings (and replacing them by a new leaf x) would result in the set $x2|34, 23|45, 34|5x$, which is incompatible.

Therefore, NJMerge does not automatically accept siblinghood proposals, since it has to check whether a given proposal creates a set of incompatible trees. The problem is that checking whether a set of unrooted trees is compatible is NP-complete [45]. As a result, NJMerge uses a polynomial time heuristic that only has the following guarantee: if it returns NO, then the siblinghood proposal definitely creates a set of incompatible trees. However, NJMerge may accept a siblinghood proposal (and hence return YES) even when the set of resulting trees would be incompatible. The problem is that NJMerge *can* (on occasion) say YES even when the siblinghood proposal should be rejected. The consequence to failing to detect incompatibility is that NJMerge will fail to return a compatibility tree on such inputs. Finally, although the incidence of failure in [30] was less than 1%, our later analyses revealed that the failure rate could be higher.

Despite having a potential to fail to return a tree (due to algorithm design, not computational issues), NJMerge has several appealing theoretical properties. The most important of these properties are the ones that provide guarantees about when NJMerge will not fail (Theorems 3) and that show that when NJMerge returns a tree it is necessarily a compatibility supertree (Theorem 4). These theorems do not assume that the set T contains leaf-disjoint trees. As a result, their conclusions apply to iterative analyses using NJMerge (where the initial input may be a set of leaf-disjoint trees, but subsequent steps in the analysis could produce sets of trees that are no longer leaf-disjoint).

Theorem 3. *Let T be a set of unrooted binary trees, leaf-labelled by elements of S, and let D be a dissimilarity matrix on S. Suppose that T is an unrooted binary tree on S with positive edge weights defining additive matrix A. Suppose also that every tree in T agrees with T, and that $\max_{ij} |D[i,j] - A[i,j]| \leq f/2$, where f is the minimum weight on any edge in T. Then NJMerge will return T on input (T, D).*

Theorem 4. *Let T be a set of unrooted binary trees, leaf-labelled by elements of S, and let D be a dissimilarity matrix on S. Suppose that NJMerge does not fail on this input, and that $NJMerge(T, D) = T$. Then T is a compatibility supertree for T.*

The proofs of these theorems are omitted but follow easily from similar theoretical results proven in [31]. These theoretical properties will be useful in Sect. 2.3, where we explore the use of NJMerge within divide-and-conquer pipelines for computing phylogenetic trees, and examine whether the pipelines are statistically consistent under different models of evolution. Finally, a big-O running time analysis of NJMerge was not provided in [30], but it is easy to see that it runs in polynomial time.

3.3 Constrained INC

Incremental Tree Building, or INC [60], is a method that was described for constructing a tree given a dissimilarity matrix D. INC is designed to be absolute fast converging (AFC) under the GTR mode (a technical term that means it has polynomial sample complexity once the length of the longest and shortest branches in the GTR model tree are bounded, as discussed in [9,10,55]). We begin by describing INC, and then describe the variant called *constrained* INC, where it can be used to build a tree given the matrix D and a set T of leaf-disjoint constraint trees.

INC has the following algorithmic design. Given D, it constructs a taxon addition ordering, $s_1, s_2, \ldots, s_n$. Then it builds a tree on the first four taxa using the Four Point Method [9], which is a simple distance-based method for constructing quartet trees. Each additional taxon is added into the growing tree

t according to the taxon addition order. When s_i is added, a linear number of quartet trees are computed (again, using the Four Point Method) and used to vote for the placement of s_i into t, and s_i is added into t using the edge with the largest number of votes. The total running time is $O(n^2)$.

Constrained INC is an extension of INC to allow the input to have the pair $\mathcal{T}$ of leaf-disjoint unrooted binary trees and a dissimilarity matrix D. It has the same basic structure as INC, modified to ensure that the constraints in $\mathcal{T}$ are obeyed. Using the sequence addition ordering (computed just as for INC), it computes a tree on the first four leaves. However, it first checks to see if all four taxa s_1, s_2, s_3, s_4 are in one of the trees in $\mathcal{T}$, and if so then it uses the induced tree on these leaves for the initialization to t; otherwise, it uses D to compute the quartet tree on s_1, s_2, s_3, s_4. Subsequent additions of taxa into the growing tree t use the constraint trees to narrow down the part of t into which each taxon can be inserted. Thus, when s_i is added into t, the edges into which s_i can be added without violating any of the trees in $\mathcal{T}$ are identified, and voting is restricted to this set. Note that it is always possible to add each subsequent taxon into t without violating the input set $\mathcal{T}$, and the final tree that is returned is a compatibility supertree for $\mathcal{T}$. A big-O running time analysis for constrained INC was not provided in [60], but since INC is $O(n^2)$, it is easy to see that constrained INC is also polynomial time. We summarize this discussion with the following theorem.

Theorem 5 *(From [60]). Let $\mathcal{T}$ be a set of unrooted binary trees, leaf-labelled by elements of S, and let D be a dissimilarity matrix on S. Then constrained INC on input $(\mathcal{T}, D)$ returns a compatibility supertree for $\mathcal{T}$, and does so in polynomial time.*

Finally, Le *et al.* [20] modified the algorithm design for INC to improve accuracy (based on simulation studies), while maintaining the theoretical guarantees (i.e., the modification to INC is still AFC, and the modification to constrained INC still satisfies the theorem above).

4   Pipelines Using DTM Methods

4.1   Overview

As seen in Fig. 1, DTM methods can be used within pipelines that use divide-and-conquer (potentially also with iteration) to improve scalability for phylogenetic estimation methods. The main approach that has been studied using DTM methods [20, 31] uses guide trees to define the decomposition, and operates as follows. The input is a set S of taxa, a method M for computing trees given subsets of S, and a DTM method Φ, and the approach operates as follows:

1. Compute a dissimilarity matrix D on the set S.
2. Construct a starting tree T.
3. Use T to compute a decomposition of S into $S_1, S_2, \ldots, S_k$ where $\cup_i S_i = S$ and $S_i \cap S_j = \emptyset$ if $i \neq j$.

4. Let $T_i = M(S_i)$ (i.e., T_i is the tree computed on subset S_i), $i = 1, 2, \ldots, k$.
5. Combine the trees $T_1, T_2, \ldots, T_k$ using Φ and D (and potentially other information about S).
6. If desired, the algorithm can repeat from Step 3, until some stopping rule is triggered. The final tree, or a consensus of the trees that are computed during the different iterations, is then returned.

Even when the phylogeny estimation method M (for computing trees on subsets) and DTM method $\dot\Phi$ are specified, there are two other algorithmic parameters that also have to be specified: how the starting tree is computed and how a tree is used to produce a division of the species set into disjoint sets. Furthermore, if iteration is used, then other decisions also need to be made to specify the algorithm. Iteration was used in [30], but even one run through the pipeline (without any iteration) improved accuracy compared to the initial tree. One challenge in using iteration is how to define the stopping rule, and also whether to return the final tree, to select from the set of trees, or to return a consensus of the trees. This is a topic for future research, and is not the focus of this paper.

Centroid Edge Decomposition. The main tree-based decomposition strategy uses the centroid edge decomposition, which was initially proposed in [22] and is used in PASTA [27], a method for co-estimating alignments and trees under sequence evolution models that include substitutions and indels (insertions and deletions). This approach uses a given tree T and target maximum subset size B, and operates as follows: Given T with leafset S, if the number of leaves in T is more than B, then find an edge e in T where deleting e (but not its endpoints) splits T into two subtrees that have close to the same number of leaves (e is called a centroid edge). Recurse on each subtree. The result is a set of disjoint subtrees of T, each with at most B leaves; the decomposition of S is obtained by returning the leaf sets of the resulting subtrees.

Computing the Starting Tree. The calculation of the starting tree T is typically done using some fast method, and prior research has shown that when iteration is used, then the general approach is fairly robust to the starting tree (as long as it is an estimated tree and not a random tree). One way of computing the starting tree is to compute a neighbor joining tree on the dissimilarity matrix D, but this initial tree can be computed in various ways, depending on the type of data. However, when the input is a set of aligned sequences, then fast maximum likelihood heuristics, such as FastTree2 [37], can be very fast and provide good starting trees. The selection of the maximum subset size B depends on the available computational resources and also on the method M, since some phylogeny estimation methods degrade in accuracy on large datasets (or have computational limitations that make analyses of large datasets infeasible).

4.2   Theoretical Guarantees

We begin with some basic theoretical properties, stating them first in the context of gene tree estimation and then in the context of species tree estimation. (Proofs are omitted, but are easily derived from similar results in [31, 60].)

Theorem 6. *Suppose that* (T, Θ) *is a GTR model tree on leafset* S, *and that* D *is the matrix of logdet distances. Let* T_0 *be any arbitrary tree on* S *and let* $S_1, S_2, \ldots, S_k$ *be pairwise-disjoint subsets of* S *with* $\cup_i S_i = S$. *Now let* M *be a method that is statistically consistent under the GTR model, let* $M(S_i) = T_i$, *and let* $\mathcal{T} = \{T_1, T_2, \ldots, T_k\}$. *Then as the sequence length increases, NJMerge($\mathcal{T}, D$) and constrained* INC($\mathcal{T}, D$) *converge to* T. *Hence the pipelines based on logdet distances and NJMerge or constrained* INC *are statistically consistent under the GTR model.*

Note that this statistical consistency guarantee does not require any particular starting tree nor particular decomposition strategy; those choices are made to improve empirical performance (and perhaps sample complexity) but are not needed for statistical guarantees.

A similar theorem can be established for multi-locus species tree estimation, but is more complicated to express.

Theorem 7. *Let* S *be a set of species, let* (T^*, Θ) *be a model species tree, and let gene trees* $T_1, T_2, \ldots, T_m$ *evolve within* T *under the MSC model. Suppose also that sequences evolve down each gene tree under the GTR model, thus producing multi-locus dataset* $\mathcal{A} = \{A_1, A_2, \ldots, A_m\}$, *where* A_i *is a multiple sequence alignment for locus* i. *Let gene trees* t_i *be estimated on each* A_i, $i = 1, 2, \ldots, m$, *using a statistically consistent method (such as maximum likelihood) and let* D *be the average internode distance matrix computed on the set of estimated gene trees. The pipeline described above, with an arbitrary starting tree, arbitrary maximum subset size, ASTRAL-III used to compute species trees on subsets, NJMerge or constrained* INC *used to combine subset trees, and dissimilarity matrix* D, *is statistically consistent under the MSC+GTR model of evolution. In other words, as the number of loci and the number of sites per locus both increase, the species tree estimated by this pipeline will converge to* T^*.

More generally, other summary methods that are statistically consistent under the MSC model (e.g., NJst and ASTRID) could also be used instead of ASTRAL-III, and produce a statistically consistent pipeline.

One of the interesting points about this theorem is that although NJMerge *can* fail on some inputs, under the conditions of the theorem (which is that the number of sites per locus and number of loci both increase), with probability converging to 1, NJMerge will not fail to return a tree when combining subset trees.

4.3   Empirical Results

Species Tree Estimation Using Divide-and-Conquer and DTM Methods. Two studies [30, 31] explored species tree estimation using divide-and-conquer pipelines with NJMerge on datasets with up to 1000 species and 1000 genes with two levels of ILS (high and low). The methods for computing species trees on subsets included SVDquartets (as implemented in PAUP* [47]), ASTRAL-III [59], and RAxML [44]. ASTRAL computes species trees from gene trees, and so gene trees were computed for each locus using FastTree2. When used with ASTRAL, D was set to be the average "internode distance matrix" (i.e., $D[i, j]$ is the average of the leaf-to-leaf topological distances in the gene trees between species i and j), and the starting tree was the neighbor joining tree on D (i.e., the starting tree is the NJst [23] tree). For the other methods (SVDquartets and RAxML), the distance matrix was the logdet distance matrix computed on the concatenated alignment and the starting tree was the neighbor joining tree on the logdet distance matrix. For the datasets with 1000 species, the maximum subset size was 120.

The results of using the pipeline with each base method (SVDquartets, ASTRAL, or RAxML) showed the following trends, compared to using the base method *de novo* (i.e., on the full dataset with up to 1000 species and 1000 genes):

- Using the pipeline with SVDquartets to compute species trees on subsets enabled it to run on 1000-species datasets (as SVDquartets cannot run *de novo* on datasets of that size), reduced running time on the smaller datasets (with 100 species), and improved accuracy compared to using SVDquartets *de novo*.
- Using the pipeline with ASTRAL to compute species trees on subsets improved running time had a minor impact on accuracy and enabled ASTRAL to complete on some datasets with 1000 species, 1000 genes, and high ILS, compared to using ASTRAL *de novo*.
- Using the pipeline with RAxML to compute species trees on subsets enabled it to complete on all 1000-species datasets (in contrast, RAxML was unable to complete on some datasets due to memory limitations when run *de novo* on datasets of that size), reduced running time, and had a minor impact on accuracy for the low ILS conditions and sometimes improved accuracy when run on the high ILS conditions, compared to running RAxML *de novo*.

Thus, the divide-and-conquer pipeline improved speed for RAxML, ASTRAL, and SVDquartets, and also enabled these methods to complete on all 1000-taxon datasets. In comparison, each method was unable to complete (either due to memory requirements or running time exceeding the limit) on some datasets. Furthermore, the divide-and-conquer pipeline improved accuracy for some methods (notably RAxML for the high ILS conditions) and otherwise had minimal impact on accuracy. These trends show that NJMerge is a valuable tool within these divide-and-conquer pipelines. However, NJMerge failed on some datasets (due to its heuristic strategy that is not guaranteed to correctly detect that siblinghood proposals are unsafe).

Gene Tree Estimation Using Divide-and-Conquer Using DTM Methods. Pipelines for gene tree estimation for use with DTM methods were explored by Le *et al.* in [20]. These pipelines are essentially the same as those studied for

NJMerge in [31], except that no iteration was explored. Thus, these pipelines begin by computing a fast starting tree, decomposing based on the centroid edge decomposition, computing trees on subsets, and then combining the trees using DTM methods. Le *et al.* studied these pipelines using NJMerge and an improvement to constrained INC they developed (as mentioned above) with constraint trees computed using maximum likelihood (ML) heuristics (RAxML and FastTree), and referred to this combination as INC-ML.

Le *et al.* explored INC-ML performance on simulated sequence datasets with up to 10,000 sequences. They also compared INC-ML to NJMerge using the same set of constraint trees on a subset of the model conditions, and observed that both DTM methods produced essentially the same accuracy for those datasets. The remaining findings, given below, were only established for INC-ML, but based on the similarity in accuracy obtained for INC-ML and NJMerge, we conjecture that these trends are likely to hold for NJMerge as well. The most important other findings in [20] were these:

- Increasing the size of the subsets computed using ML heuristics improved accuracy but increased running time; correspondingly, using INC alone (i.e., not within the divide-and-conquer pipeline) was much less accurate than INC-ML.
- The starting tree had little impact on the final accuracy of INC-ML.
- The choice of ML heuristic (RAxML or FastTree) within INC-ML had an impact on the final accuracy, but neither method clearly outperformed the other.
- INC-ML was consistently more accurate than neighbor joining and less accurate than RAxML, and usually (but not always) less accurate than FastTree. Furthermore, often the differences were not small.

The last of these observations is obviously disappointing.

4.4   Discussion

Results for gene tree estimation are shown for divide-and-conquer pipelines using two DTM methods (NJMerge and constrained INC), with trees computed on subsets with two maximum likelihood heuristics, RAxML and FastTree2. Notably, the trees computed using divide-and-conquer pipelines with these two DTM methods had indistinguishable accuracy. However, results here are discouraging, as accuracy using the divide-and-conquer approach was reduced compared to using the better of the two ML heuristics, RAxML, applied to the entire dataset. The reduction in accuracy is likely to reflect both the outstanding suitability of the ML criterion for gene tree estimation (since there is a perfect fit between the model that generated the input sequence data and the model assumed by the phylogeny estimation method) as well as the RAxML search heuristic. If so, the only benefit of divide-and-conquer strategies for ML gene tree estimation would be computational: enabling the better ML methods to run on larger datasets or to complete in less time or using fewer resources.

The results here for species tree estimation are limited to NJMerge, but show that divide-and-conquer combined with NJMerge can provide large benefits. Specifically, we showed that the centroid edge decomposition strategy used with ASTRAL, a leading species tree estimation method, showed improvements in speed but no significant change in accuracy. When used with RAxML (a concatenation analysis that is not statistically consistent for species tree estimation as it does not take gene tree heterogeneity into account), the divide-and-conquer strategy sometimes improved accuracy and otherwise had minimal impact on accuracy; the pipeline also improved running time and enabled RAxML to run on some datasets where it was unable to run *de novo* due to memory limitations. Finally, SVDquartets was unable to run *de novo* on any 1000-taxon datasets but the pipeline enabled it to run on all the datasets.

For both gene tree and species tree estimation, improvements in speed occur when the methods used to construct trees on subsets are much more computationally intensive than the methods used to combine subset trees. However, the divide-and-conquer strategy reduced accuracy in the case of gene tree estimation but either improved accuracy or produced a very modest decrease in accuracy for species tree estimation. Understanding why divide-and-conquer has a difference in impact is important in order to understand the conditions under which this type of strategy should be helpful.

Recall that the gene tree estimation context involved constructing trees on sequences that have evolved down GTR model trees, using maximum likelihood heuristics that assume the GTR model. In other words, for the gene tree estimation problem there is a perfect fit between the data generation model and the tree estimation model. In contrast, for the species tree estimation problem the methods used to estimate species trees are not as perfectly suited to the data generation process. For example, RAxML assumes all sites evolve down a single GTR model tree, which is not true in the multi-locus setting in the presence of gene tree heterogeneity. ASTRAL is statistically consistent under the MSC and so will converge to the true species tree as the number of true gene trees increases; however, the input to ASTRAL is a set of estimated gene trees rather than a set of true gene trees, so that even in this case there is model misspecification. Finally, SVDquartets, despite its positive theoretical properties, has not been shown to provide comparable accuracy to ASTRAL [32,53]. In other words, all the species tree estimation methods we examined have properties that suggest there is room for improvement in terms of accuracy.

5   Conclusions

This paper presented a simple divide-and-conquer strategy that can be used with phylogeny estimation methods, towards the goal of improving scalability to large datasets, reducing running time, and potentially improving empirical accuracy. The results discussed here show improvements in speed and scalability for species tree estimation but not for gene tree estimation, suggesting some potential limitations for this approach. However, changes in the approach, including the development of new DTM methods or new approaches for dividing a dataset into disjoint sets, might provide better empirical accuracy, as well as maintaining theoretical guarantees. Thus, future research may result in new divide-and-conquer methods that provide improvements in scalability, possibly even when used with RAxML for gene tree estimation.

The established theory is also quite limited. For example, INC-NJ (which used a different decomposition technique for defining the subsets for computing constraint trees) is AFC for GTR gene tree estimation, but other divide-and-conquer approaches based on DTM methods may also be provably AFC. Furthermore, although INC-NJ had poor accuracy (compared to other methods) in [20], we conjecture that new divide-and-conquer strategies could improve performance on simulated data for base methods, much as DCM1 [55] improved both the theoretical sample complexity and empirical performance for neighbor joining and other exponentially converging methods [33]. Future research should explore this, and should also explore the impact on sample complexity for species tree estimation [43].

Finally, other types of phylogeny estimation problems could benefit from these or future divide-and-conquer strategies, such as species tree estimation in the presence of gene duplication and loss [4,5,50] or gene tree estimation in the presence of heterotachy [17,24,25,49,61]. More generally, this paper is meant as a starting point for future investigations, in the hope that subsequent research will yield improved methods and software that can lead to methods capable of reconstructing highly accurate trees on ultra-large datasets.

Acknowledgments. This work was supported in part by NSF grant CCF-1535977. I also wish to thank Erin Molloy and Thien Le for helpful comments on the manuscript.

References

1. Atteson, K.: The performance of neighbor-joining methods of phylogenetic reconstruction. Algorithmica **25**, 251–278 (1999)
2. Bansal, M., Burleigh, J., Eulenstein, O., Fernández-Baca, D.: Robinson-Foulds supertrees. Algorithms Mol. Biol. **5**, 18 (2010)
3. Bayzid, M.S., Hunt, T., Warnow, T.: Disk covering methods improve phylogenomic analyses. BMC Genom. **15**(Suppl. 6), S7 (2014)
4. Boussau, B., Szöllősi, G., Duret, L., Gouy, M., Tannier, E., Daubin, V.: Genome-scale co-estimation of species and gene trees. Genom. Res. **23**, 323–330 (2013)
5. Chaudhary, R., Bansal, M.S., Wehe, A., Fernández-Baca, D., Eulenstein, O.: iGTP: a software package for large-scale gene tree parsimony analysis. BMC Bioinform. **11**, 574 (2010)
6. Chaudhary, R., Burleigh, J.G., Fernández-Baca, D.: Fast local search for unrooted Robinson-Foulds supertrees. IEEE/ACM Trans. Comput. Biol. Bioinform. **9**, 1004–1013 (2012)
7. Chifman, J., Kubatko, L.: Quartet inference from SNP data under the coalescent. Bioinformatics **30**(23), 3317–3324 (2014)
8. Chifman, J., Kubatko, L.: Identifiability of the unrooted species tree topology under the coalescent model with time-reversible substitution processes, site-specific rate variation, and invariable sites. J. Theor. Biol. **374**, 35–47 (2015)

9. Erdös, P., Steel, M., Székely, L., Warnow, T.: A few logs suffice to build (almost) all trees (i). Random Struct. Algorithms **14**, 153–184 (1999)
10. Erdös, P., Steel, M., Székely, L., Warnow, T.: A few logs suffice to build (almost) all trees (ii). Theor. Comput. Sci. **221**, 77–118 (1999)
11. Felsenstein, J.: Evolutionary trees from DNA sequences: a maximum likelihood approach. J. Mol. Evol. **17**(6), 368–376 (1981)
12. Felsenstein, J.: Inferring Phylogenies. Sinauer Associates, Sunderland (2004)
13. Heled, J., Drummond, A.J.: Bayesian inference of species trees from multilocus data. Mol. Biol. Evol. **27**, 570–580 (2010)
14. Huson, D., Nettles, S., Warnow, T.: Disk-covering, a fast converging method for phylogenetic tree reconstruction. J. Comput. Biol. **6**(3), 369–386 (1999)
15. Jukes, T.H., Cantor, C.R.: Evolution of protein molecules. In: Mammalian Protein Metabolism, pp. 21–132 (1969)
16. Kingman, J.F.C.: The coalescent. Stochast. Process. Appl. **13**, 235–248 (1982)
17. Kolaczkowski, B., Thornton, J.: Performance of maximum parsimony and likelihood phylogenetics when evolution is heterogeneous. Nature **431**, 980–984 (2004). https://doi.org/10.1038/nature02917
18. Kubatko, L., Degnan, J.: Inconsistency of phylogenetic estimates from concatenated data under coalescence. Syst. Biol. **56**, 17 (2007)
19. Lagergren, J.: Combining polynomial running time and fast convergence for the disk-covering method. J. Comput. Syst. Sci. **65**(3), 481–493 (2002)
20. Le, T., Sy, A., Molloy, E., Zhang, Q., Rao, S., Warnow, T.: Using INC within divide-and-conquer phylogeny estimation. In: Proceedings of AlCoB 2019 (2019)
21. Lefort, V., Desper, R., Gascuel, O.: FastME 2.0: a comprehensive, accurate, and fast distance-based phylogeny inference program. Mol. Biol. Evol. **32**(10), 2798–2800 (2015). https://doi.org/10.1093/molbev/msv150
22. Liu, K., et al.: SATé-II: very fast and accurate simultaneous estimation of multiple sequence alignments and phylogenetic trees. Syst. Biol. **61**(1), 90–106 (2012). https://doi.org/10.1093/sysbio/syr095
23. Liu, L., Yu, L.: Estimating species trees from unrooted gene trees. Syst. Biol. **60**(5), 661–667 (2011)
24. Lockhart, P., Novis, P., Milligan, B., Riden, J., Rambaut, A., Larkum, T.: Heterotachy and tree building: a case study with Plastids and Eubacteria. Mol. Biol. Evol. **23**(1), 40–45 (2006) https://doi.org/10.1093/molbev/msj005. http://mbe.oxfordjournals.org/content/23/1/40.abstract
25. Lopez, P., Casane, D., Philippe, H.: Heterotachy, an important process of protein evolution. Mol. Biol. Evol. **19**, 1–7 (2002)
26. Maddison, W.: Gene trees in species trees. Syst. Biol. **46**(3), 523–536 (1997)
27. Mirarab, S., Nguyen, N., Wang, L.S., Guo, S., Kim, J., Warnow, T.: PASTA: ultra-large multiple sequence alignment of nucleotide and amino acid sequences. J. Comput. Biol. **22**, 377–386 (2015)
28. Mirarab, S., Reaz, R., Bayzid, M.S., Zimmermann, T., Swenson, M., Warnow, T.: ASTRAL: accurate species TRee ALgorithm. Bioinformatics **30**(17), i541–i548 (2014)
29. Mirarab, S., Warnow, T.: ASTRAL-II: coalescent-based species tree estimation with many hundreds of taxa and thousands of genes. Bioinformatics **31**(12), i44–i52 (2015)
30. Molloy, E.K., Warnow, T.: NJMerge: a generic technique for scaling phylogeny estimation methods and its application to species trees. In: Blanchette, M., Ouangraoua, A. (eds.) RECOMB-CG 2018. LNCS, vol. 11183, pp. 260–276. Springer, Cham (2018). https://doi.org/10.1007/978-3-030-00834-5_15

31. Molloy, E.K., Warnow, T.: Statistically consistent divide-and-conquer pipelines for phylogeny estimation using NJMerge. bioRxiv (2018). https://doi.org/10.1101/469130

32. Molloy, E.K., Warnow, T.: To include or not to include: the impact of gene filtering on species tree estimation methods. Syst. Biol. **67**(2), 285–303 (2018). https://doi.org/10.1093/sysbio/syx077

33. Nakhleh, L., Roshan, U., St. John, K., Sun, J., Warnow, T.: Designing fast converging phylogenetic methods. Bioinformatics **17**, 190–198 (2001)

34. Nelesen, S., Liu, K., Wang, L.S., Linder, C.R., Warnow, T.: DACTAL: divide-and-conquer trees (almost) without alignments. Bioinformatics **28**, i274–i282 (2012)

35. Nguyen, L.T., Schmidt, H., von Haeseler, A., Minh, B.: IQ-TREE: a fast and effective stochastic algorithm for estimating maximum-likelihood phylogenies. Mol. Biol. Evol. **32**(1), 268–274 (2015). https://doi.org/10.1093/molbev/msu300

36. Nguyen, N., Mirarab, S., Warnow, T.: MRL and SuperFine+ MRL: new supertree methods. Algorithms Mol. Biol. **7**(1), 3 (2012)

37. Price, M., Dehal, P., Arkin, A.: FastTree 2 - approximately maximum-likelihood trees for large alignments. PLoS ONE **5**(3), e9490 (2010). https://doi.org/10.1371/journal.pone.0009490

38. Roch, S.: A short proof that phylogenetic tree reconstruction by maximum likelihood is hard. TCBB **3**(1), 92–94 (2006)

39. Roch, S., Nute, M., Warnow, T.: Long-branch attraction in species tree estimation: inconsistency of partitioned likelihood and topology-based summary methods. Syst. Biol. **68**, 281–297 (2018). https://doi.org/10.1093/sysbio/syy061

40. Roch, S., Steel, M.: Likelihood-based tree reconstruction on a concatenation of aligned sequence data sets can be statistically inconsistent. Theor. Popul. Biol. **100**, 56–62 (2015)

41. Ronquist, F.: Matrix representation of trees, redundancy, and weighting. Syst. Biol. **45**, 247–253 (1996)

42. Saitou, N., Nei, M.: The neighbor-joining method: a new method for reconstructing phylogenetic trees. Mol. Biol. Evol. **4**, 406–425 (1987)

43. Shekhar, S., Roch, S., Mirarab, S.: Species tree estimation using ASTRAL: how many genes are enough? IEEE/ACM Trans. Comput. Biol. Bioinform. (TCBB) **15**(5), 1738–1747 (2018)

44. Stamatakis, A.: RAxML-VI-HPC: maximum likelihood-based phylogenetic analyses with thousands of taxa and mixed models. Bioinformatics **22**, 2688–2690 (2006)

45. Steel, M.: The complexity of reconstructing trees from qualitative characters and subtrees. J. Classif. **9**, 91–116 (1992)

46. Steel, M.: Recovering a tree from the leaf colourations it generates under a Markov model. Appl. Math. Lett. **7**, 19–24 (1994)

47. Swofford, D.L.: PAUP*. Phylogenetic analysis using parsimony (*and other methods). Version 4. Sinauer Associates (2003)

48. Tavaré, S.: Some probabilistic and statistical problems in the analysis of DNA sequences. Lect. Math. Life Sci. **17**, 57–86 (1986). American Mathematical Society

49. Taylor, M., Kai, C., Kawai, J., Carninci, P., Hayashizaki, Y., Semple, C.: Heterotachy in mammalian promoter evolution. PLoS Genet. **2**(4), e30 (2006). https://doi.org/10.1371/journal.pgen.0020030

50. Ullah, I., Parviainen, P., Lagergren, J.: Species tree inference using a mixture model. Mol. Biol. Evol. **32**(9), 2469–2482 (2015)

51. Vachaspati, P., Warnow, T.: ASTRID: accurate species TRees from internode distances. BMC Genom. **16**(Suppl. 10), S3 (2015)

52. Vachaspati, P., Warnow, T.: FastRFS: fast and accurate Robinson-Foulds supertrees using constrained exact optimization. Bioinformatics (2016). https://doi.org/10.1093/bioinformatics/btw600

53. Vachaspati, P., Warnow, T.: SVDquest: improving SVDquartets species tree estimation using exact optimization within a constrained search space. Mol. Phylogenet. Evol. **124**, 122–136 (2018). https://doi.org/10.1016/j.ympev.2018.03.006

54. Wang, L.S., Leebens-Mack, J., Wall, P.K., Beckmann, K., DePamphilis, C.W., Warnow, T.: The impact of multiple protein sequence alignment on phylogenetic estimation. IEEE/ACM Trans. Comput. Biol. Bioinform. **8**, 1108–1119 (2011)

55. Warnow, T., Moret, B.M.E., St. John, K.: Absolute convergence: true trees from short sequences. In: Proceedings of ACM-SIAM Symposium on Discrete Algorithms (SODA 2001), pp. 186–195. Society for Industrial and Applied Mathematics (SIAM), Philadelphia (2001)

56. Warnow, T.: Computational Phylogenetics: An Introduction to Designing Methods for Phylogeny Estimation. Cambridge University Press, Cambridge (2017)

57. Warnow, T.: Divide-and-conquer tree estimation: opportunities and challenges. In: Warnow, T. (ed.) Bioinformatics and Phylogenetics. Springer, Heidelberg (2019)

58. Yang, Z.: Molecular Evolution: A Statistical Approach. Oxford University Press, Oxford (2014)

59. Zhang, C., Sayyari, E., Mirarab, S.: ASTRAL-III: Increased scalability and impacts of contracting low support branches. In: Meidanis, J., Nakhleh, L. (eds.) RECOMB-CG 2017. LNCS, pp. 53–75. Springer, Cham (2017). https://doi.org/10.1007/978-3-319-67979-2_4

60. Zhang, Q., Rao, S., Warnow, T.: New absolute fast converging phylogeny estimation methods with improved scalability and accuracy. In: Parida, L., Ukkonen, E. (eds.) 18th International Workshop on Algorithms in Bioinformatics (WABI 2018), pp. 8:1–8:12. LIPICS, Dagsttuhl (2018)

61. Zhou, Y., Rodrigue, N., Lartillot, N., Philippe, H.: Evaluation of the models handling heterotachy in phylogenetic inference. BMC Evol. Biol. **7**, 206 (2007)

62. Zimmermann, T., Mirarab, S., Warnow, T.: BBCA: improving the scalability of *BEAST using random binning. BMC Genom. **15**(Suppl. 6), S11 (2014). Proceedings of RECOMB-CG (Comparative Genomics)

Biological Networks and Graph Algorithms

New Polynomial-Time Algorithm Around the Scaffolding Problem

Tom Davot[1]($\boxtimes$)(iD), Annie Chateau[1,2](iD), Rodolphe Giroudeau[1](iD),
and Mathias Weller[3](iD)

[1] LIRMM, CNRS UMR 5506, Montpellier, France
{tom.davot,annie.chateau,rodolphe.giroudeau}@lirmm.fr
[2] IBC, Montpellier, France
[3] LIGM, Bât Copernic, Champs-sur-Marne, France
mathias.weller@u-pem.fr

Abstract. We describe in this paper an approximation algorithm for the scaffolding problem, which is part of genome inference in bioinformatics. The aim of the problem is to find a maximum weighted collection of disjoint alternating cycles and paths covering a particular graph called scaffold graph. The problem is known to be NP-complete, and we describe further result concerning a special class of graphs aiming to be close to real instances. The described algorithm is the first polynomial-time approximation algorithm designed for this problem on non-complete graphs.

Keywords: Genome scaffolding · Approximation ·
Dynamic programming

1 Introduction

Motivation. We are interested here in an algorithmic problem occurring in the production of genomes. Genomes are usually obtained by sequencing, which produces a set of *reads* whose length and quality depend on the sequencing technology. It is commonly known that short reads (typically hundreds of base pairs), produced by second generation sequencing technology (Illumina), are of better quality than long reads (thousands of base pairs), produced by third generation sequencing technologies (PacBio or Oxford Nanopore) [6]. Those reads are then assembled using a variety of tools, the most recent integrating very efficient hybrid strategies using both short and long reads [7]. However, databases are full of "old genomes", produced before the development of third generation sequencing, and "hard genomes" that escape sequencing in good conditions. In fact, most of the genomes in databases consist as huge sets of chunks of sequences, called *contigs*. These sets contain far more contigs than the real number of the chromosomes of the corresponding organisms. Such fragmentation is observed even for well-studied genomes. To the natural question "how to reduce this fragmentation?", technological progress and costly re-sequencing is not the only answer and computational exploitation of already available sequencing data is possible.

© Springer Nature Switzerland AG 2019
I. Holmes et al. (Eds.): AlCoB 2019, LNBI 11488, pp. 25–38, 2019.
https://doi.org/10.1007/978-3-030-18174-1_2

Scaffolding Problem. We focus here on the contig scaffolding problem which, given a set of contigs, asks to infer the order and the orientation of the contigs along the target genome, using a set of possibly inconsistent pairing information. This information could be provided, for instance, by paired-end reads whose two ends map to distinct contigs. Formally, it is possible to extract from this information a set of relationships between the contigs, that may be inconsistent. A survey on recent methods is available in [5]. The problem that we model here is more general than those presented in the literature, and therefore allows adaptation towards realistic modelization. We study the scaffolding problem as an optimization problem in a graph called *scaffold graph*, obtained by mapping of paired-end reads on *de novo* contigs. However, the present formulation is not limited to this aspect, and may also consider other sources of information. We consider that the scaffolding may obey genomic structural constraints, like a fixed number of linear and circular chromosomes. In the past, we presented preliminary results about the complexity of this problem and a first polynomial-time approximation algorithm on cliques [1]. Those results were extended and completed by another polynomial-time approximation algorithm [2], and by a randomized approach [3]. Exact approaches have been explored [9], leading to study sparse cases [10]. The contribution of the present paper is a continuation of [4,8], where special classes of graphs has been studied, from sparse to very dense. Real instances are very sparse, but show some dense regions. Hence, we are interested in graphs built from cliques separated by bridges (i.e. edges whose removal disconnects the graph).

In the following, most of the proofs has been omitted due to space constraints. A full version including the proofs is available in https://hal-lirmm.ccsd.cnrs.fr/lirmm-02047701.

2   Notation and Problem Description

In this section, we formally define the SCAFFOLDING problem. For a graph G, we denote by $V(G)$ and $E(G)$ the set of vertices and edges of G, respectively. A *scaffold graph* (G^*, M^*, ω) is a simple loopless graph G^* with a perfect matching M^* and a weight function ω on the non-matching edges. The matching M^* represents the set of contigs and the function ω represents the confidence that two contigs occur consecutively in the genomic sequence. An *alternating path* (resp. *alternating cycle*) is a path (resp. cycle) such that its edges alternatively belong to M^* or not. The extremal edges of an alternating path must be in M^*. The SCAFFOLDING problem, whose decision version is NP-complete on general graphs [2], is defined as follows:

(σ_p, σ_c)-SCAFFOLDING (SCA)
Input: a scaffold graph (G^*, M^*, ω) and integers σ_p, σ_c.
Task: Find a collection S of σ_p alternating paths and σ_c alternating cycles
maximizing $\sum_{e \in S \setminus M^*} \omega(e)$

The two integers σ_p and σ_c are used to model the genomic structure by representing the number of linear and circular chromosomes, respectively. Let

---

Algorithm 1. Polynomial-time approximation algorithm for (σ_p, σ_c)-
SCAFFOLDING PROBLEM.

Data: A scaffold graph (G^*, M^*, ω), two integers σ_p and σ_c.
Result: A collection of σ_p alternating paths and σ_c alternating cycles or
  "False" if no such collection exists.

    // Initialization step

1 $S \leftarrow M^*$;

2 $E \leftarrow E \setminus M^*$;

3 sort E by decreasing order of weight;

4 if not $Feasibility((G^*, M^*), S, \sigma_p, \sigma_c)$ then return False;

    // Main loop

5 while $E \neq \varnothing$ do

6   Let $e = \{u, v\}$ be the first element in the ordered-list E;

7     $E \leftarrow E \setminus e$;

8   if $Feasiblity((G^*, M^*), S \cup \{e\}, \sigma_p, \sigma_c)$ then

9        $R \leftarrow$ set of edges of E incident to e;

10       $S \leftarrow S \cup \{e\}$;

11      $E \leftarrow E \setminus R$;

12 return S;

S be a partial solution of SCAFFOLDING, the *cardinality* of S is the number of alternating paths and cycles which compose S. We denote by $\sigma_p(S)$ and $\sigma_c(S)$ the number of alternating paths and alternating cycles of S, respectively. The approximation algorithm used in the following is described in Algorithm 1. It is known that this algorithm produces a solution for SCAFFOLDING with an approximation ratio of three in complete graphs [2].

To adapt Algorithm 1 for another class of graphs, we need to provide a dedicated feasibility function. This function, given a partial solution S, indicates if it is possible to build a solution to SCAFFOLDING with the remaining edges. In this paper, we use Algorithm 1 on a particular class of graphs defined as follows:

Definition 1. *A connected cluster graph G is a graph which admits a decomposition of its edges $E(G) = E' \cup B$ such that the subgraph induced by E' is a disjoint union of cliques and each edge $e \in B$ is a bridge of G.*

An example of a connected cluster graph is given in Fig. 1. Let G be a connected cluster graph, for sake of simplicity, we designate by *clique* a connected component of the subgraph induced by E' and we denote by $CC(G)$ the set of cliques of G.

As the structure of a connected cluster graph is close to a tree (that is, shrinking each clique of G^* into a single vertex leads to a tree), we use a similar vocabulary: a *rooted* connected cluster graph is a connected cluster graph where a clique r is designated as a *root* and in that case, the *parent* of a clique is the clique connected to it on the path to r. A *child* of a clique c is the clique of which c is the parent. A vertex v of a clique c is a *door* of c if a child of c is adjacent to v. The *upper door* of c is the vertex adjacent to the parent of c. In the following, we will focus on scaffold graphs (G^*, M^*, ω) such that G^* is a connected cluster graph and $M^* \cap B = \varnothing$. Since the decision version of SCAFFOLDING is NP-complete on cliques [2], it is also NP-complete on connected cluster graphs.

3 Feasibility

In this section, we present an algorithm to determine if it is possible to construct a solution of SCAFFOLD-ING in a connected cluster graph. Its principle is to construct and assemble some partial solutions in a bottom-up traversal of the connected cluster graph. Instead of storing the feasible solutions, we store their cardinalities.

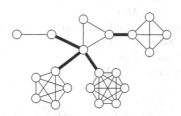

Fig. 1. Example of connected cluster graph. The bridge edges are bold.

Operations. Let G_1 and G_2 be two edge-disjoint subgraphs. We can build a solution in the graph induced by $V(G_1) \cup V(G_2)$, from a solution in G_1 and a solution in G_2, using four operations.

Definition 2. *Let G_1 and G_2 be edge-disjoint subgraphs of G^*. Let S_1 and S_2 be solutions of G_1 and G_2, respectively. Let S be a solution of $G^*[V(G_1) \cup V(G_2)]$. S is a composition of S_1 and S_2 if exactly one of the following operations occurs:*

Merger: *merge a path of S_1 with a path of S_2 in S.*
Closing: *close a path of S_1 and a path of S_2 into an alternating cycle in S.*
Absorption: *replace a non-matching edge vv' of S_2 by an alternating path of S_1, that is, $S = S_1 \cup (S_2 \setminus \{vv'\}) \cup \{uv, u'v'\}$ where u and u' are extremities of a u-u'-path in S_1. We call vv' absorbent.*
Juxtaposition: *S is the disjoint union of S_1 and S_2 and none of the previous operations are performed.*

To implement these operations, we add edges of $E(G^*) \setminus (E(G_1) \cup E(G_2))$ to S.

Note that a composition of two solutions does not always exist, except for the juxtaposition operation. In the algorithm, we manipulate sets of solutions instead of solutions. Thus, we can create a new set of solutions if all the solutions of the two input sets are used in the resulting set.

Definition 3. *Let G_1 and G_2 be two edge-disjoint subgraphs of G^* and let $\mathcal{S}_1$ and $\mathcal{S}_2$ be sets of solutions of subgraphs G_1 and G_2, respectively. Then, we call the set $\mathcal{S} = \{S \mid \exists S_1 \in \mathcal{S}_1, \exists S_2 \in \mathcal{S}_2 \text{ s.t. } S \text{ is a composition of } S_1 \text{ and } S_2\}$ the complete composition of $\mathcal{S}_1$ and $\mathcal{S}_2$.*

To ensure the possibility of building a complete composition from two sets of solutions, it is useful to characterize a solution according to the operations we can perform on it. Thus, given two subgraphs G_1 and G_2, we define four properties on a solution S according to the operations on S.

Definition 4. *Let G and G' be two vertex-disjoint subgraphs of G^* and let S be a feasible solution of* SCAFFOLDING *for (G, M^*, ω).*

1. *We call S closeable if S contains an alternating u-v-path and there is an alternating[1] u-v-path in $G' \cup \{u, v\}$.*
2. *We call S extensible by G' if S contains a vertex v such that v is an extremity of an alternating path and v has a neighbor in G' .*
3. *We call S frozen to G' if S is not extensible.*
4. *We call S absorbent to G' if S contains a non-matching edge uv and G' contains a matching edge $u'v'$ such that $uu', vv' \in E(G^*)$.*

For simplicity, we sometimes omit to precise G' and in this case $G' = G^* - V(G)$. Note that all closeable solutions are also extensible. If a solution S is closeable by a subgraph G', then we can close an alternating path of S into an alternating cycle by adding some edges of G'. If a solution S is extensible by a subgraph G', then we can add some edges of G' in an extremity of an alternating path of S without changing the cardinality of the solution. Finally, if a solution S is absorbent to a subgraph G', then we can replace an absorbent edge of S by a path of length three without changing the cardinality of S. An example of the different operations of Definition 4 is given in Fig. 2.

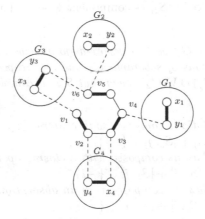

Fig. 2. The solution S is composed of a single alternating path $\{v_1, \ldots, v_6\}$. S is closeable by subgraph $G_3 = \{x_3, y_3\}$: we can close the alternating path of S into an alternating cycle by adding the edges v_1x_3, x_3y_3 and y_3v_6. S is extensible by subgraph $G_2 = \{x_2, y_2\}$: we can extend the alternating path of S by adding the edges v_6y_2 and y_2x_2 without changing the number of paths in S. S is absorbent to $G_4 = \{x_4, y_4\}$: we can replace the edge v_2v_3 of S by the edges v_2y_4, y_4x_4 and x_4v_3 without changing the number of paths in S. S is frozen to $G_1 = \{x_1, y_1\}$.

---

[1] We use here "alternating" in an abusive manner, meaning alternating matching edges and non-matching edges, beginning and ending with non-matching edges.

Semantics. Since the number of possible solutions can be exponential, we just store the possible cardinalities in the table entries, which is sufficient to answer the question of feasibility. We recall that, if $X, Y \subseteq \mathbb{N}$ are two sets of integers, then the sum of X and Y is defined as $X + Y = \{x + y \mid x \in X, y \in Y\}$. Note that $X + \varnothing = \varnothing$.

Definition 5. *Let $\mathcal{S}$ be a set of solutions and $i, j \in \mathbb{N}$. Then, j is called* eligible *with respect to $(\mathcal{S}, i)$ if there is a solution $S \in \mathcal{S}$ containing i alternating cycles and j alternating paths.*

Our dynamic programming table has the following semantics.

Semantics. *Let $\mathcal{S}$ be a set of solutions and $i \in \mathbb{N}$. A table entry $[\mathcal{S}, i]$ is the set of all integers eligible with respect to the tuple $(\mathcal{S}, i)$. More formally, letting $\mathcal{S}_i = \{S \mid S \in \mathcal{S} \wedge \sigma_c(S) = i\}$, we define $[\mathcal{S}, i] = \bigcup_{S \in \mathcal{S}_i} \{\sigma_p(S)\}$.*

Let us highlight three particular values of $[\mathcal{S}, i]$. For $\mathcal{S} = \{\varnothing\}$, we have $[\{\varnothing\}, 0] = 0$ and, for each $i > 0$, we have $[\{\varnothing\}, i] = \varnothing$. For an alternating path p, we have $[\{p\}, 0] = 1$ and $[\{p\}, i] = \varnothing$ for each $i > 0$. Finally, for an alternating cycle c, we have $[\{c\}, 1] = 0$ and $[\{c\}, i] = \varnothing$ for each $i \neq 1$. For simplicity, we denote by $[\mathcal{S}]$ the vector $([\mathcal{S}, 0], \ldots, [\mathcal{S}, \sigma_c])$ and, for any operator $\diamond$ and any sets $\mathcal{S}_1$ and $\mathcal{S}_2$ of solutions, we define $[\mathcal{S}_1] \diamond [\mathcal{S}_2]$ as component-wise $\diamond$, that is, $[\mathcal{S}_1, i] \diamond [\mathcal{S}_2, i]$ for each $i \in [0, \sigma_c]$.

Lemma 6. *Let G_1 and G_2 be two vertex-disjoint subgraphs of G^* and let $\mathcal{S}_1$ and $\mathcal{S}_2$ be sets of solutions of subgraphs G_1 and G_2, respectively. Let $\mathcal{S}$ be a set of solutions of $G^*[V(G_1) \cup V(G_2)]$ such that $\mathcal{S}$ is a complete composition of $\mathcal{S}_1$ and $\mathcal{S}_2$.*

1. *If $\mathcal{S}$ is the set of solutions composed with a merger operation, then*
   $\forall i, j, [\mathcal{S}, i + j] = [\mathcal{S}_1, i] + [\mathcal{S}_2, j] + \{-1\}$.
2. *If $\mathcal{S}$ is the set of solutions composed with a closing operation, then*
   $\forall i, j, [\mathcal{S}, i + j + 1] = [\mathcal{S}_1, i] + [\mathcal{S}_2, j] + \{-2\}$.
3. *If $\mathcal{S}$ is the set of solutions composed with an absorption operation, then*
   $\forall i, j, [\mathcal{S}, i + j + 1] = [\mathcal{S}_1, i] + [\mathcal{S}_2, j] + \{-1\}$.
4. *If $\mathcal{S}$ is the set of solutions composed with a juxtaposition operation, then*
   $\forall i, j, [\mathcal{S}, i + j + 1] = [\mathcal{S}_1, i] + [\mathcal{S}_2, j]$.

We use Lemma 6 to define four applications *juxtapose*, *merge$_t$*, *absorb*, *close$_t$* which provide table entries for complete compositions "composed" with a juxtaposition, merger, absorption or closing operation, respectively. Although Lemma 6 is defined for two sets, we use a generalized version which can take as parameters more than two sets. The functions *merge$_t$* and *close$_t$* have a parameter t which indicates the number of paths merged or closed during the operation. For example, if we have three sets $\mathcal{S}_1$, $\mathcal{S}_2$, and $\mathcal{S}_3$ and it is possible to construct a single alternating path in the resulting composition by taking one alternating path in each set, then we use the function *merge$_3$*$(\{\mathcal{S}_1\}, \{\mathcal{S}_2\}, \{\mathcal{S}_3\})$. In addition, it is sometimes possible to close a single alternating path into an alternating cycle and in that case the function $close_1$ is used. The four applications are defined in Algorithms 7, 8 and Algorithm 9 (in appendix). However, we must ensure that the associated operation is feasible before using one this application.

The Algorithm. We now present a method to provide the feasibility function needed by Algorithm 1. We suppose that a partial solution S is given. Let c be a clique of G^* and let S' be the intersection of S and c. An *alternating element* of c is either an alternating cycle of S' or an alternating path of S'. We traverse different types of subgraphs defined in the following way:

- Let $v \in V(G^*)$, let $C(v)$ be the set of children adjacent to v (possibly empty). The subgraph $G^*(v)$ is the union of v and all branches incident to v. Formally, $G^*(v) = G^*[\{v\} \cup \bigcup_{c \in C(v)} V(G^*(c))]$.
- Let e be an alternating element, the subgraph $G^*(e)$ is the union of e and all children incident to one of its vertices. Formally, $G^*(e) = G^*[\bigcup_{v \in e} V(G^*(v))]$.
- Let c be a clique of G^*, and let dd' be the matching edge of c incident to the upper door of c. Let $c' = c \setminus dd'$ be the *subclique* of c. For all $x \in \{c, c'\}$, the subgraph $G^*(x)$ is the union of x and all children incident to a vertex of x. Formally, $G^*(x) = G^*[\bigcup_{e \in M^*(x)} V(G^*(e))]$.

For each traversed subgraph G', we use four different sets of solutions distinguishing solutions according to their properties.

Definition 7. *Let S be a partial solution of G^*. Let x be a vertex, a partial path, a subclique or clique of G^* and let S' be a solution of the subgraph $G^*(x)$.*

- *$S \in C(x) \Leftrightarrow S'$ is closeable and $S \cap G^*(x) \in S'$.*
- *$S \in P(x) \Leftrightarrow S \notin C(x)$ and S is extensible and $S \cap G^*(x) \in S'$.*
- *$S \in A(x) \Leftrightarrow S$ is frozen and absorbent and $S \cap G^*(x) \in S'$.*
- *$S \in F(x) \Leftrightarrow S \notin A(x)$ and S is frozen and $S \cap G^*(x) \in S'$.*

The next paragraphs are dedicated to describing the algorithms to calculate the table entries for the four types of subgraphs described above.

Vertex. Let v be a vertex of G^*. We show in this part how to compute the table entries for the sets $F(v)$ and $P(v)$. Note that, since the edge between $G^*(v)$ and its parent is a bridge, the sets $C(v)$ and $A(v)$ are empty. Any solution S' of $G^*(v)$ can have at most one incident edge to v. If no edge of $S \cap G^*(v)$ is incident to v, the idea is to construct the table entries by merging successively the table entries of the children incident to v. For that, we use at each step an intermediate graph G_i. Let V_i be the union of the i first children of v. G_i is the subgraph of G^* induced by v and all vertices in V_i. Otherwise, if one edge of $S \cap G^*(v)$ is incident to v, then any solution containing S belongs to $P(v)$.

Lemma 8. *For any vertex v, the values of the table entries provided by Algorithm 2 are correct for the set $F(v)$ and $P(v)$.*

Algorithm 2. *compute_vertex*

Data: A scaffold graph (G^*, M^*), a partial solution S and a vertex v.

1   $[\mathcal{F}(v)] \leftarrow \varnothing$; $[\mathcal{P}(v)] \leftarrow \varnothing$; $[\mathcal{F}(v), 0] \leftarrow \{0\}$;
2   $C \leftarrow \{c_1, \ldots, c_k\}$: list of children linked to v;
3   **foreach** $c_t \in C(v)$ **do**
4      $compute\_clique(c_t)$;
5      $[\mathcal{F}'] \leftarrow [\mathcal{F}(v)]$;
6      $[\mathcal{P}'] \leftarrow [\mathcal{P}(v)]$;
7   **if** $\exists uv \in E(G^*(v)) \cap S$ **then**
8   **if** $u \in c_t$ **then**
9          $[\mathcal{P}(v)] \leftarrow juxtapose(\{\mathcal{P}'\}, \{\mathcal{P}(c_t)\})$
10   **else**
11          $[\mathcal{P}(v)] \leftarrow juxtapose(\{\mathcal{P}'\}, \{\mathcal{F}(c_t), \mathcal{P}(c_t)\})$
12   **else**
13        $[\mathcal{F}(v)] \leftarrow juxtapose(\{\mathcal{F}'\}, \{\mathcal{F}(c_t), \mathcal{P}(c_t)\})$
14        $[\mathcal{P}(v)] \leftarrow juxtapose(\{\mathcal{P}'\}, \{\mathcal{F}(c_t), \mathcal{P}(c_t)\})$
           $\cup \quad juxtapose(\{\mathcal{F}'\}, \{\mathcal{P}(c_t)\})$

Alternating Element. Let c be a clique of G^* and e be an alternating element of c such that e does not contain the upper door of c. We show in this part how to compute the table entries for the sets $\mathcal{C}(e), \mathcal{F}(e)$ and $\mathcal{P}(e)$. If e is a $u-v$-path, then the idea is to merge the computed table entries of u and v and juxtapose the frozen solutions of the inner vertices. If e is an alternating cycle, then there is no choice to do and the only solution containing S is frozen.

Lemma 9. *For any alternating element e, the values of the table entries provided by Algorithm 3 are correct for the sets $\mathcal{C}(e), \mathcal{F}(e)$ and $\mathcal{P}(e)$.*

Note that the only possibility to obtain an absorbent solution of $G^*(e)$ is when e is a path and become closed into an alternating cycle. However, suppose that an absorption operation is done in the function *compute_subclique*. The resulting solution can also be obtained by a closing operation with a solution in $\mathcal{C}(e)$. Thus, to avoid recurrence, the value of $[\mathcal{A}(e)]$ is not provided.

Subclique. Let c' be a subclique of G^*. We show in this part how to compute the table entries for the sets $\mathcal{C}, \mathcal{F}, \mathcal{A}$ and $\mathcal{P}$. The idea is to construct the table entry by merging successively each table entry of the alternating elements of c'. For that, we use at each step an intermediate graph G_i and three intermediate sets $\mathcal{F}_+, \mathcal{A}_+$ and $\mathcal{P}_+$. Let E_i be the i first alternating element of c' and $V_i = \bigcup_{e \in E_i} V(G^*(e))$. G_i is the subgraph of G^* induced by V_i. At step i, a solution $S' \in \mathcal{F}_+$ if and only if (1) S' is a solution of G_i, (2) S' contains a set $C \neq \varnothing$ of closeable paths and (3) $S \setminus C$ is frozen. The sets $\mathcal{A}_+$ and $\mathcal{P}_+$ are defined similarly (only the condition (3) changes).

Algorithm 3. compute_alternating_element

Data: A scaffold graph (G^*, M^*), a partial solution S and an alternating element e with vertices $\{v_0, v_1, \ldots, v_k\}$.

1 **foreach** $v \in p$ **do** *compute_vertex*(v) ;
2 **if** e *is an alternating cycle* **then**
3     $[\mathcal{F}(e)] \leftarrow$ *juxtapose*$(\{e\}, \{\mathcal{F}(v_0)\}, \ldots, \{\mathcal{F}(v_k)\})$;
4     $[\mathcal{C}(e)] \leftarrow \varnothing; [\mathcal{A}(e)] \leftarrow \varnothing; [\mathcal{P}(e)] \leftarrow \varnothing$;

5 **else**
6     $[\mathcal{I}_e] \leftarrow$ *juxtapose*$(\{\mathcal{F}(v_1)\}, \ldots, \{\mathcal{F}(v_{k-1})\})$;
7     $[\mathcal{C}(e)] \leftarrow$ *juxtapose*$(\{e\}, \{\mathcal{F}(v_0)\}, \{\mathcal{F}(v_k)\}, \{\mathcal{I}_e\})$;
8     $[\mathcal{F}(e)] \leftarrow$ *merge₃*$(\{e\}, \{\mathcal{P}(v_0)\}, \{\mathcal{P}_v(v_k)\}, \{\mathcal{I}_e\})$;
        $\cup$ *close₁*$(\{e\}, \{\mathcal{F}(v_0)\}, \{\mathcal{F}(v_k)\}, \{\mathcal{I}_e\})$;
9     $[\mathcal{P}(e)] \leftarrow$ *merge₂*$(\{e\}, \{\mathcal{P}(u)\}, \{\mathcal{F}(v)\}, \{\mathcal{I}_e\})$;
        $\cup$ *merge₂*$(\{e\}, \{\mathcal{F}(u)\}, \{\mathcal{P}(v)\}, \{\mathcal{I}_e\})$;

Lemma 10. *For any subclique c', the value of the table entries provided by Algorithm 4 are correct for the sets $\mathcal{C}(c'), \mathcal{F}(c'), \mathcal{A}(c')$ and $\mathcal{P}(c')$.*

Clique. Let c be a clique of G^* and d be the upper door of c. We show in this part how to compute the table entries for the sets $\mathcal{F}(c)$ and $\mathcal{P}(c)$. Note that since the edge between $G^*(c)$ and its parent is a bridge, the sets $\mathcal{C}(c)$ and $\mathcal{A}(c)$ are empty. Let e be the alternating element of c containing the upper door d. The idea is to first compute the table entries for the graph $G^*(e)$ and then merge the obtained table entries to the table entries of the subclique. If e is an alternating path and d is an extremity of e, we replace $\mathcal{P}(e)$ by two intermediate sets $\mathcal{P}_d$ and $\mathcal{P}_{d'}$. Let S' be a solution of $G^*(e)$. $S' \in \mathcal{P}_d$ if and only if $S' \in \mathcal{P}(e)$ and d is an extremity of an alternating path of S'. Likewise, $S' \in \mathcal{P}_{d'}$ if and only if $S' \in \mathcal{P}(e)$ and d is not an extremity of an alternating path of S'. Note that $\mathcal{P}(e) = \mathcal{P}_d \cup \mathcal{P}_{d'}$. In order to compute these two sets, we reuse the value of $\mathcal{I}_e$, computed in *compute_alternating_element*.

Lemma 11. *For any clique c, the values of the table entries provided by Algorithm 5 are correct for the sets $\mathcal{F}(c)$ and $\mathcal{P}(d)$.*

3.1 Feasability Function

Finally, we can now provide an answer to the feasibility of finding a solution for SCAFFOLDING problem by using Algorithm 6.

Corollary 12. *Given a partial solution S, Algorithm 6 returns true if and only if (G^*, M^*) can be decomposed into σ_p alternating paths and σ_c alternating cycles. The time complexity of the algorithm is $\mathcal{O}(|V(G^*)| \cdot \sigma_c^2)$.*

Proof. Since $G^*(root) = G^*$, it exists a solution S with $\sigma_p(S) = \sigma_p$ and $\sigma_c(S) = \sigma_c$, if and only if S belongs to $\mathcal{C}(root) \cup \mathcal{F}(root), \cup \mathcal{A}(root) \cup \mathcal{P}(root)$.

Algorithm 4. *compute_subclique*

Data: A scaffold graph (G^*, M^*), a partial solution S and a subclique c'.

1   $[\mathcal{F}(c')] \leftarrow \varnothing; [\mathcal{P}(c')] \leftarrow \varnothing; [\mathcal{A}(c')] \leftarrow \varnothing;$

2   $[\mathcal{F}_+] \leftarrow \varnothing; [\mathcal{P}_+] \leftarrow \varnothing; [\mathcal{A}_+] \leftarrow \varnothing;$

3   $[\mathcal{F}(c'), 0] \leftarrow \{0\};$

4   $E \leftarrow \{e_1, \ldots, e_k\}$: list of alternating elements of c';

5   **foreach** $e_t \in E$ **do**

6      $compute\_alternating\_element(e_t);$

7      $[\mathcal{F}'] \leftarrow [\mathcal{F}(c')]; [\mathcal{P}'] \leftarrow [\mathcal{P}(c')]; [\mathcal{A}'] \leftarrow [\mathcal{A}(c')];$

8      $[\mathcal{F}'_+] \leftarrow [\mathcal{F}_+]; [\mathcal{P}'_+] \leftarrow [\mathcal{P}_+]; [\mathcal{A}'_+] \leftarrow [\mathcal{A}_+];$

9

10      $[\mathcal{F}(c')] \leftarrow juxtapose(\{\mathcal{F}'\}, \{\mathcal{F}(e_t)\})$

11

12      $[\mathcal{F}_+] \leftarrow juxtapose(\{\mathcal{F}'_+\}, \{\mathcal{F}(e_t)\})$
               $\cup \;\; juxtapose(\{\mathcal{F}', \mathcal{F}_+\}, \{\mathcal{C}(e_t)\})$

13

14      $[\mathcal{A}(c')] \leftarrow juxtapose(\{\mathcal{A}'\}, \{\mathcal{F}(e_t)\})$
               $\cup \;\; merge_2(\{\mathcal{P}'\}, \{\mathcal{P}(e_t)\})$
               $\cup \;\; absorb(\{\mathcal{A}'\}, \{\mathcal{C}(e_t)\})$
               $\cup \;\; close_2(\{\mathcal{F}'_+, \mathcal{A}'_+\}, \{\mathcal{C}(e_t)\})$

15

16      $[\mathcal{A}_+] \leftarrow juxtapose(\{\mathcal{A}, \mathcal{A}'_+\}, \{\mathcal{F}(e_t), \mathcal{C}(e_t)\})$
               $\cup \;\; merge_2(\{\mathcal{P}'_+\}, \{\mathcal{P}(e_t)\})$
               $\cup \;\; merge_2(\{\mathcal{F}'_+, \mathcal{A}'_+\}, \{\mathcal{C}(e_t)\})$

17

18      $[\mathcal{P}(c')] \leftarrow juxtapose(\{\mathcal{P}'\}, \{\mathcal{F}(e_t), \mathcal{P}(e_t)\})$
               $\cup \;\; juxtapose(\{\mathcal{F}', \mathcal{A}'\}, \{\mathcal{P}(e_t)\})$
               $\cup \;\; merge_2(\{\mathcal{F}'_+\}, \{\mathcal{P}(e_t)\})$
               $\cup \;\; merge_2(\{\mathcal{P}'\}, \{\mathcal{C}(e_t)\})$
               $\cup \;\; close_2(\{\mathcal{P}'_+\}, \{\mathcal{C}(e_t)\})$

19

20      $[\mathcal{P}_+] \leftarrow juxtapose(\{\mathcal{P}'_+\}, \{\mathcal{F}(e_t), \mathcal{C}(e_t)\})$
               $\cup \;\; juxtapose(\{\mathcal{F}'_+, \mathcal{A}'_+, \}, \{\mathcal{P}(e_t)\})$
               $\cup \;\; juxtapose(\{\mathcal{P}'\}, \{\mathcal{C}(e_t)\})$

21

22 **end**

23 $[\mathcal{C}(c')] \leftarrow [\mathcal{F}_+] \cup [\mathcal{A}_+] \cup [\mathcal{P}_+]$

Thus the return of the function indicates if such a solution exists and then the algorithm is correct. Concerning the time complexity, the composition operations are in $(O)(\sigma_c^2)$. Thus, without taking in account the reccursive calls, the time complexity of Algorithms 2, 3, 4, 5 in one iteration of a loop is $\mathcal{O}(\sigma_c^2)$. In Algorithm 2, the number of iterations made by all calls of this function depends on $|CC(G^*)|$ G^* and then the time complexity of all this iterations is $(O)(|CC(G^*)| \cdot \sigma_c^2)$. Similary we can show that the time complexity of the iterations made by all calls of Algorithm 3, 4, 5 are $(O)(|V| \cdot \sigma_c^2)$, $(O)(|M^*| \cdot \sigma_c^2)$ and

---

Algorithm 5. *compute_clique*

Data: A scaffold graph (G^*, M^*), a partial solution S and a clique c.

1   $d \leftarrow$ upper door of c; $e \leftarrow$ alternating element of c containing d;

2   *compute_subclique*(c'); *compute_alternating_element*(e);

3   **if** e *is an alternating path* **and** d *is an extremity of* e **then**

4      $d' \leftarrow$ other extremity of e;

5      $[\mathcal{P}_d] \quad\leftarrow\; juxtapose(\{e\}, \{\mathcal{F}(d)\}, \{\mathcal{P}(d')\}, \{\mathcal{I}_e\})$

6      $[\mathcal{P}_{d'}] \quad\leftarrow\; juxtapose(\{e\}, \{\mathcal{P}(d)\}, \{\mathcal{F}(d')\}, \{\mathcal{I}_e\})$

7

8      $[\mathcal{F}(c)] \quad\leftarrow\; juxtapose(\{\mathcal{F}(e), \mathcal{P}_{d'}\}, \{\mathcal{F}(c'), \mathcal{C}(c'), \mathcal{A}(c'), \mathcal{P}(c')\})$
            $\cup \quad merge_2(\{\mathcal{C}(e), \mathcal{P}_{d'}\}, \{\mathcal{C}(c'), \mathcal{P}(c')\})$
            $\cup \quad absorb(\{\mathcal{C}(e)\}, \{\mathcal{A}(c')\})$
            $\cup \quad close_2(\{\mathcal{C}(e)\}, \{\mathcal{C}(c')\})$

9

10     $[\mathcal{P}(c)] \quad\leftarrow\; juxtapose(\{\mathcal{C}(e), \mathcal{P}_d\}, \{\mathcal{F}(c'), \mathcal{C}(c'), \mathcal{A}(c'), \mathcal{P}(c')\})\}$
            $\cup \quad merge_2(\{\mathcal{C}(e)\}, \{\mathcal{C}(c'), \mathcal{P}(c')\})$

11 **end**

12 **else**

13     $[\mathcal{F}(c)] \quad\leftarrow\; juxtapose(\{\mathcal{C}(c'), \mathcal{F}(c'), \mathcal{A}(c'), \mathcal{P}(c')\}, \{\mathcal{C}(e), \mathcal{F}(e), \mathcal{P}(e)\})$
            $\cup \quad merge_2(\{\mathcal{C}(c'), \mathcal{P}(c')\}, \{\mathcal{C}(e), \mathcal{P}(e)\})$
            $\cup \quad absorb_1(\{\mathcal{A}(c')\}, \{\mathcal{C}(e)\})$
            $\cup \quad close(\{\mathcal{C}(c')\}, \{\mathcal{C}(e)\})$

14     $[\mathcal{P}(c)] \quad\leftarrow\; \varnothing$

15   ;

16 **end**

Algorithm 6. *Feasibility*

Data: A scaffold graph (G^*, M^*) a partial solution S and two integers σ_p, σ_c

1   $root \leftarrow$ root of G^* ;

2   *compute_subclique*$(root)$;

3   **return** $\sigma_p \in ([\mathcal{C}(root), \sigma_c] \cup [\mathcal{F}(root), \sigma_r] \cup [\mathcal{A}(root), \sigma_c] \sqcup [\mathcal{P}(root), \sigma_c])$

$(O)(|CC(G^*)| \cdot \sigma_c^2)$. Then, the time complexity of all iterations in all function is $(O)((|V| + |M^*| + |CC(G^*)|) \cdot \sigma_c^2)$ and since the number of matching edges and the number of cliques is bounded by the number of vertices of G^*, we have a time complexity in $\mathcal{O}(|V(G^*)| \cdot \sigma_c^2)$.

4   Approximation Result

4.1   Notations and Definitions

The algorithm presented in this section is an adaptation of the one described in [2]. The original algorithm works in complete graph and we adapt it so that it works in connected cluster graph. Let (G^*, M^*, ω) be a scaffold connected cluster graph. The idea of the algorithm is to visit each non matching edge of (G^*, M^*, ω) by decreasing order of weight and chose some of them to be part of the solution S. We start by running Algorithm 6 on (G^*, M^*, ω). This run allows us to both verify if a solution is feasible and to initialize the different table entries. When an edge uv is visited, it can be added in S or removed from (G^*, M^*, ω). When an edge uv is chosen, then all non-matching edges incident to u or v are removed from the list of sorted edges. At each step, we must ensure that we can build a solution with the remaining edges, that is, uv is added in S if and only if we can build σ_p alternating paths and σ_c alternating cycles with the remaining edges.

4.2 Optimization

At each step, we must run the Algorithm 6 in order to check if it possible to construct a solution. However, it is not necessary to update all the table entries at each step. In fact, when an edge uv is tested, it is only necessary to update the table entries of the cliques containing u or v and then, update all the table entries of the ancestors of this cliques.

4.3 Algorithm

Lemma 13. *Algorithm 1 provides a solution for the (σ_p, σ_c)-SCAFFOLDING in path connected cluster graph with an approximation ratio of five and a time complexity $\mathcal{O}(|V| \cdot |E(G^*)| \cdot \sigma_c^2)$. The approximation ratio is tight.*

5 Conclusion

We presented in this paper the first polynomial-time algorithm approximating the scaffolding problem on non-complete graphs. Using a dynamic programming approach, we exploited the tree-like nature of connected cluster graph to extend the feasibility function and the analysis of the approximation ratio. A natural extension of this work would be to explore the practical aspects of this algorithm. Since connected cluster graphs aim to model real instances, we intend to measure in what extend this algorithm provides good results on them. We expect the ratio on real instances to be close to one, as for the greedy algorithm on cliques [1]. We may also explore the possibility to exploit randomized algorithms framework to improve this ratio.

A   Appendix

A.1   Algorithms

Algorithm 7. *juxtapose*

Data: $\mathcal{S}^1 = \{\mathcal{S}_1^1, \mathcal{S}_2^1, \dots\}, \dots, \mathcal{S}^k = \{\mathcal{S}_1^k, \mathcal{S}_2^k, \dots\}$: sets of sets of solutions.

1 **if** $k = 0$ **then**
2 | $[\mathcal{S}] \leftarrow 0$;
3 **end**
4 $[\mathcal{I}] \leftarrow juxtapose(S^2, \dots, S^k)$;
5 **forall** $i \in [0, \sigma_c]$ **do**
6 | **forall** $j \in [0, \sigma_c - i]$ **do**
7 | | $[\mathcal{S}, i+j] \leftarrow [\mathcal{S}, i] + \bigcup_{S \in \mathcal{S}^1}[S, j]$
8 | **end**
9 **end**
10 **return** $[\mathcal{S}]$

Algorithm 8. *merge$_t$ or absorb*

Data: $\mathcal{S}^1 = \{\mathcal{S}_1^1, \mathcal{S}_2^1, \dots\}, \dots, \mathcal{S}^k = \{\mathcal{S}_1^k, \mathcal{S}_2^k, \dots\}$: sets of sets of solutions,
      t: number of paths to merge ($t = 2$ in the absorb function).

1 **forall** $i \in [0, \sigma_c]$ **do**
2 | **forall** $j \in [0, \sigma_c - i]$ **do**
3 | | $[\mathcal{S}, i+j] \leftarrow \bigcup_{S \in \mathcal{S}^1}[S, i] + \bigcup_{S' \in \mathcal{S}^2}[S', j] + \{-(t-1)\}$
4 | **end**
5 **end**
6 **if** $k \neq 2$ **then**
7 | $[\mathcal{S}] \leftarrow juxtapose(\{S\}, S^3, \dots, S^k)$;
8 **end**
9 **return** $[\mathcal{S}]$

Algorithm 9. *close$_t$*

Data: $\mathcal{S}^1 = \{\mathcal{S}_1^1, \mathcal{S}_2^1, \dots\}, \dots, \mathcal{S}^k = \{\mathcal{S}_1^k, \mathcal{S}_2^k, \dots\}$: sets of sets of solutions,
      t: number of paths to close.

1 **forall** $i \in [0, \sigma_c]$ **do**
2 | **forall** $j \in [0, \sigma_c - i]$ **do**
3 | | $[\mathcal{S}, i+j+1] \leftarrow \bigcup_{S \in \mathcal{S}^1}[S, i] + \bigcup_{S' \in \mathcal{S}^2}[S', j] + \{-t\}$
4 | **end**
5 **end**
6 **if** $k \neq 2$ **then**
7 | $[\mathcal{S}] \leftarrow juxtapose(\{S\}, S^3, \dots, S^k)$;
8 **end**
9 **return** $[\mathcal{S}]$

References

1. Chateau, A., Giroudeau, R.: Complexity and polynomial-time approximation algorithms around the scaffolding problem. In: Dediu, A.-H., Martín-Vide, C., Truthe, B. (eds.) AlCoB 2014. LNCS, vol. 8542, pp. 47–58. Springer, Cham (2014). https://doi.org/10.1007/978-3-319-07953-0_4
2. Chateau, A., Giroudeau, R.: A complexity and approximation framework for the maximization scaffolding problem. Theor. Comput. Sci. **595**, 92–106 (2015)
3. Chen, Z., Harada, Y., Machida, E., Guo, F., Wang, L.: Better approximation algorithms for scaffolding problems. In: Zhu, D., Bereg, S. (eds.) FAW 2016. LNCS, pp. 17–28. Springer, Cham (2016). https://doi.org/10.1007/978-3-319-39817-4_3
4. Dallard, C., Weller, M., Chateau, A., Giroudeau, R.: Instance guaranteed ratio on greedy heuristic for genome scaffolding. In: Chan, T.-H.H., Li, M., Wang, L. (eds.) COCOA 2016. LNCS, vol. 10043, pp. 294–308. Springer, Cham (2016). https://doi.org/10.1007/978-3-319-48749-6_22
5. Mandric, I., Lindsay, J., Măndoiu, I.I., Zelikovsky, A.: Scaffolding algorithms (chap. 5). In: Măndoiu, I., Zelikovsky, A. (eds.) Computational Methods for Next Generation Sequencing Data Analysis. Wiley, Hoboken (2016)
6. Mardis, E.R.: DNA sequencing technologies: 2006–2016. Nat. Protoc. **12**(2), 213–218 (2017)
7. Miller, J.R., et al.: Hybrid assembly with long and short reads improves discovery of gene family expansions. BMC Genomics **18**(1), 541 (2017)
8. Weller, M., Chateau, A., Dallard, C., Giroudeau, R.: Scaffolding problems revisited: complexity, approximation and fixed parameter tractable algorithms, and some special cases. Algorithmica **80**(6), 1771–1803 (2018). https://doi.org/10.1007/s00453-018-0405-x
9. Weller, M., Chateau, A., Giroudeau, R.: Exact approaches for scaffolding. BMC Bioinform. **16**(Suppl. 14), S2 (2015)
10. Weller, M., Chateau, A., Giroudeau, R.: On the complexity of scaffolding problems: from cliques to sparse graphs. In: Lu, Z., Kim, D., Wu, W., Li, W., Du, D.-Z. (eds.) COCOA 2015. LNCS, vol. 9486, pp. 409–423. Springer, Cham (2015). https://doi.org/10.1007/978-3-319-26626-8_30

Enumerating Dominant Pathways in Biological Networks by Information Flow Analysis

Ozan Kahramanoğulları[✉][iD]

Department of Mathematics, University of Trento, Trento, Italy
ozan.kah@gmail.com

Abstract. Cells perceive and respond to their microenvironment as a part of their functioning via networks of processes resulting from molecular interactions. The complexity of such networks has been the subject of studies that address their various aspects. Some of these include static methods that focus on graph representations and their consequent properties, while others take a dynamical systems approach based on simulations. Here, we address the problem of identifying dominant pathways in biological networks that are represented as activation and repression edges. For this purpose, we propose a hybrid method that combines static graph properties with a dynamic quantification of information flow that results from stochastic simulations. We first illustrate our method on a simple example, and then apply it to the *Escherichia coli* transcription network consisting of 4639 regulatory edges.

Keywords: Dominant pathways · Biological networks · Information flow analysis

1 Introduction

The complex biological processes dedicated to sustaining life are commonly represented by various kinds of networks for formal analysis. These representations cover a broad spectrum from detailed rule-based models and chemical reaction networks to other stoichiometric representations as well as graph models with varying details. These different methods have their individual strengths in delivering new insights on a rich landscape of biological queries. However, more than often, the availability of empirical data, or the lack thereof, poses a bottleneck in the formal setting within the context of specific studies.

Despite the inherent complexity, experimental findings indicate that certain aspects and patterns are common in many biological networks. Some of these features resemble those in the networks that are observed outside the biochemical realm such as communication networks. By relying on these similarities, here we work with the consideration of networks of biochemical entities as processes that relay the incoming stimulus to response components. We aim at benefitting

© Springer Nature Switzerland AG 2019
I. Holmes et al. (Eds.): AlCoB 2019, LNBI 11488, pp. 39–51, 2019.
https://doi.org/10.1007/978-3-030-18174-1_3

from this notion especially when larger biological processes are considered and kinetic data is scarce or difficult to apply. In particular, we address the problem of identifying dominant pathways: within an organism, biological processes work together to produce an overall global flux distribution by making the flow of information accessible over different pathways [10,12,16,20]. Despite the presence of many pathways, some of the available pathways are more dominant in terms of their information flow capacity.

In the light of the observations above, our method is based on the following assumptions on biological systems. Firstly, information propagates by a series of coupled biochemical reactions, whereby cellular signals are relayed predominantly through functional modules of highly connected nodes, see, e.g., [13]. In particular, in the extreme case of scale-free networks, whose degree distribution follows a power law [1], only a tiny fragment of hub nodes process a large fragment of the information. Secondly, local signal transduction tends to be stochastic, hence information propagation by individual components is subject to noise. And finally, in the presence of multiple pathways for the signal, information flow has a predilection for the pathway of least biochemical resistance.

Our analysis of biological systems combines methods from static graph theoretical considerations in the literature with those for dynamical systems based on simulations. We work with biological systems that can be represented as directed graphs with two kinds of edges, namely activation and repression. Many biological systems can be represented in this form as well as gene transcription networks that easily fit into this category. We map each edge of the network to a reaction of a chemical reaction network (CRN). The idea here is that each activation edge consumes the instance of the incoming signal at its source node to propagate the information flow by producing an instance of the target node. The signal can then be passed on to the next reaction. Each repression edge, on the other hand, consumes the instance of the incoming signal together with an instance of its target node if it is available. This way, it inhibits the further propagation of the signal from its target node.

In accordance with the considerations above, we resort to the idea that cellular signals are transmitted dominantly through pathways of highly connected components. We implement our method by computing the reaction rates as proxies of connectivity of the source and target nodes of the corresponding edge in the network. We instantiate the rates by borrowing three topological measures from the literature that are used to study the static properties of graphs. These are the *topological overlap measure* (TOM) [18], the Randić index (RI) [17], and the combined linkage index (CLI) [15]. For each network, we produce three different CRNs using these measures to instantiate the reaction rates, together with a control CRN that assigns 1.0 to all the rates.

The dynamic component of our method is realised by running stochastic simulations on the CRNs by using the Gillespie algorithm that implements mass action kinetics [5]. In previous work, we have developed a conservative extension of this algorithm that traces species fluxes during simulation [8,9]. We compare the results with TOM, RI, CLI models and the control model by quantifying the distance between their results. We quantify the flow of information in different models in terms of their simulation fluxes on the mean of repeated simulations. We then apply exhaustive breadth-first search on the resulting flux graphs to enumerate the pathways and rank them.

In the following, we illustrate our method on a simple network. We then apply it to the *E. coli* transcription network [4] with 4639 regulatory edges.[1]

2   Information Flow in Biological Networks

2.1   Network Implementation

We work with networks consisting of directed graphs with two kinds of edges. Formally, an information flow network $\mathcal{G} = (\mathcal{V}, \mathcal{A}, \mathcal{I})$ is given with

- the set $\mathcal{V}$ of vertices representing the biochemical molecules or events;
- the set $\mathcal{A}$ of directed activation edges where the source activates the target;
- and the set $\mathcal{I}$ of directed inhibition edges where the source inhibits the target.

Example 1. The network depicted in Fig. 1 provides a description of a fragment of the dopamine signalling network.

The degree of a node x, denoted with $deg(x)$, is the number of edges incident to the node, with loops counted twice. For two nodes, we define $edge(x, y)$ as the number of edges, be it activation or inhibition, from x to y. We define $int(x, y)$ as the number of nodes that are connected with a single edge to both x and y.

Example 2. In the network depicted in Fig. 1, $deg(GBetaGamma) = 6$ and $deg(PQCaCh) = 2$. We have that $edge(\mathsf{GBetaGama}, \mathsf{PQCaCh}) = 1$. Because they do not have any common neighbours, $int(\mathsf{GBetaGamma}, \mathsf{PQCaCh}) = 0$.

At the first step, our algorithm for computing the information flow maps the network to a chemical reaction network (CRN), whereby activation and inhibition edges are given with two different kinds of reactions.

The activation edges of the form (x, y) are mapped to reactions

$$\mathsf{x} \xrightarrow{r} \mathsf{y},$$

which model the information flow from x to y, and r is the rate of the reaction. By relying on the notion that cellular signals are transmitted dominantly through pathways of highly connected components, we compute the reaction rates as proxies of connectivity of the source and target nodes of the corresponding edge in the network. For this, we employ three different measures from the literature.

---

[1] All the data and scripts are available for download at: ozan-k.com/pathways.zip.

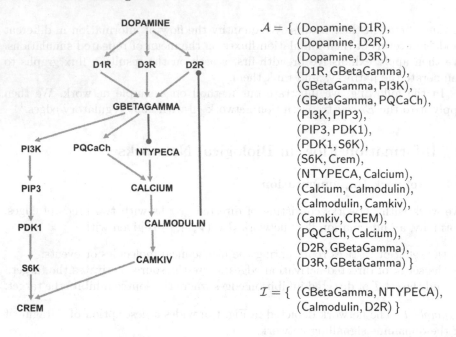

DOPAMINE

D1R   D3R   D2R

GBETAGAMMA

PI3K   PQCaCh   NTYPECA

PIP3   CALCIUM

PDK1   CALMODULIN

S6K   CAMKIV

CREM

$\mathcal{A} = \{$ (Dopamine, D1R),
(Dopamine, D2R),
(Dopamine, D3R),
(D1R, GBetaGamma),
(GBetaGamma, PI3K),
(GBetaGamma, PQCaCh),
(PI3K, PIP3),
(PIP3, PDK1),
(PDK1, S6K),
(S6K, Crem),
(NTYPECA, Calcium),
(Calcium, Calmodulin),
(Calmodulin, Camkiv),
(Camkiv, CREM),
(PQCaCh, Calcium),
(D2R, GBetaGamma),
(D3R, GBetaGamma) $\}$

$\mathcal{I} = \{$ (GBetaGamma, NTYPECA),
(Calmodulin, D2R) $\}$

Fig. 1. The network given with $\mathcal{G} = (\mathcal{V}, \mathcal{A}, \mathcal{I})$, where $\mathcal{V} = \{$Dopamine, D1R, GBetaGama, PI3K, PIP3, PDK1, S6K, CREM, NTYPECA, Calcium, Calmodulin, Camkiv, PQCaCh, D2R, D3R$\}$, together with its graphical representation, whereby inhibitory edges are depicted in red and with round arrowheads. (Color figure online)

1. The *topological overlap measure* (TOM) [18], which was originally introduced to study the relationship between the network structure and the functional organisation of cellular metabolisms. We obtain the TOM rate value r as:

$$r = \frac{int(x,y)+1}{min(deg(x),\ deg(y))}$$

2. The Randić index (RI) [17] has been related to physical and chemical properties of organic molecules. We apply it to a single edge as follows:

$$r = \frac{1}{\sqrt{deg(x).deg(y)}}$$

3. The combined linkage index (CLI) [15] extends RI with the aim of emphasising the strongest links of each node.

$$r = \frac{edge(x,y)+edge(y,x)+2.int(x,y)}{\sqrt{(deg(x)+1).(deg(y)+1)}},$$

The inhibitory edges of the form (x, y) are mapped to reactions of the form

$$x+y \xrightarrow{r'} \cdot.$$

Such a reaction models the consumption of the information at x to inhibit the further downstream flow from y to other components of the network. The reaction rate is given by $r' = r.p$, whereby r is defined as above and the constant p is the inhibition constant, which factors for these second order reactions that can have a much higher propensity in comparison to first-order activation reactions. In our analysis, we first use a default value of 1.0 for p, and also evaluate the effect of smaller values.

Example 3. By applying the definitions of TOM, RI and CLI to the network in Fig. 1, we obtain the CRN with the reactions listed in Fig. 2 and together with the rate values listed in Table 1.

2.2 Network Simulation

CRNs can be simulated stochastically by using Gillespie's direct method, which is also known as the stochastic simulation algorithm (SSA) [5]. Various extensions of SSA in the literature address a variety of concerns such as increasing efficiency of simulations, simulation of rare events or others, e.g., [3,6,11]. In previous work [8,9], we have presented a method that extends SSA for stochastic flux analysis of CRNs. The method, called fSSA, is a conservative extension of SSA that monitors the distribution of the network resources during simulation with respect to the causal interdependence of the reaction instances. This consideration originates from non-interleaving models of concurrent computations used in computer science [7,14]. In such a setting, the dependencies are observed in a manner that takes into account the propensity of each reaction in terms of the resources available to that reaction. As a result of this, simulations

Table 1. The CRN obtained from the network depicted in Fig. 1, and its r values according to TOM, RI and CLI. In the simulations, we have varied the inhibitory constant p between 10^{-4} and 1.0. The resulting flux graph is depicted in Fig. 2 and normalised flux values for different p values are listed in Table 6.

Reactions	TOM	RI	CLI	Reactions	TOM	RI	CLI
1	0.5	0.41	0.29	11	0.5	0.5	0.33
2	0.33	0.33	0.25	12	0.5	0.41	0.29
3	0.5	0.41	0.29	13	0.33	0.33	0.25
4	0.5	0.29	0.22	14	0.5	0.41	0.29
5	0.5	0.29	0.22	15	$0.33.p$	$0.33.p$	$0.25.p$
6	$0.22.p$	$0.29.p$	$0.44.p$	16	0.5	0.5	0.33
7	0.5	0.29	0.22	17	0.5	0.41	0.29
8	0.5	0.5	0.33	18	0.33	0.24	0.19
9	0.5	0.5	0.33	19	0.5	0.29	0.22
10	0.5	0.5	0.33				

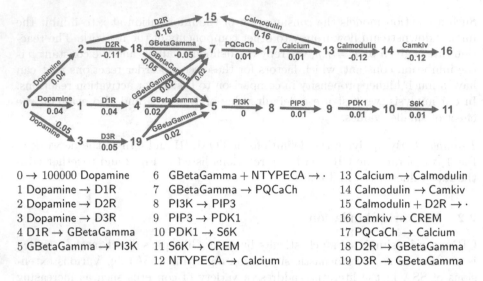

0 → 100000 Dopamine	6 GBetaGamma + NTYPECA → ·	13 Calcium → Calmodulin
1 Dopamine → D1R	7 GBetaGamma → PQCaCh	14 Calmodulin → Camkiv
2 Dopamine → D2R	8 PI3K → PIP3	15 Calmodulin + D2R → ·
3 Dopamine → D3R	9 PIP3 → PDK1	16 Camkiv → CREM
4 D1R → GBetaGamma	10 PDK1 → S6K	17 PQCaCh → Calcium
5 GBetaGamma → PI3K	11 S6K → CREM	18 D2R → GBetaGamma
	12 NTYPECA → Calcium	19 D3R → GBetaGamma

Fig. 2. The fluxes of the network in Fig. 1, delivered by the simulations with the CRN above and the rates listed in Table 1. The simulations are initiated with 100000 Dopamine as the system input. The node numbers are the CRN reactions. The inhibition reaction 15 is indicated with an underline. The complete flux data with TOM, RI and CLI is given in Table 6. The numbers on the edges summarise the data: each number denotes the maximum difference in normalised flux resulting from increasing the inhibition constant from $p = 10^{-4}$ to $p = 1.0$ in all cases.

resulting from our algorithm provide a quantitative view of the flow of information in the network besides the usual time series information. The flux graphs, that are output by the algorithm, reflect what fragment of system resources flow through which pathways of the network. This kind of information becomes particularly significant when a system resource is produced or consumed by multiple components. In this regard, flux graphs display which components produce and consume such resources. For example, in the network above, GBetaGamma production and consumption can follow many different pathways in the network.

We use the fSSA algorithm to run simulations on the CRNs. During these simulations, flux graphs can be obtained for the whole simulation interval as well as for arbitrary time intervals. In contrast to similar considerations with ordinary differential equations, these time intervals can be transient intervals, whereby the system has not yet reached its steady state levels, given by the ordinary differential equation simulations. Because the flow of resources can take different pathways at different intervals of the simulation, such a capability is essential for analysing the system behaviour at different stages.

For the example network in Fig. 1, we have obtained the flux graph depicted in Fig. 2. The measures described above, that is, TOM, RI and CLI, result in different reaction rates, listed in Table 1, thus they result in different values for the fluxes. However they all result in the same topology depicted in Figs. 2 and 3.

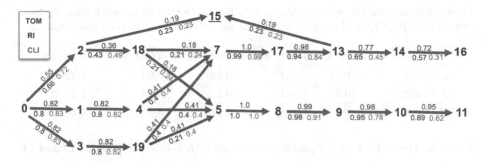

Fig. 3. The fluxes in Fig. 2 with TOM (red), RI (blue) and CLI (green) measures. (Color figure online)

Due to stochasticity, each simulation with the same CRN produces slightly different fluxes. For a systematic comparison that takes into account these variations as well as the effect of the different measures, we have first set a control network, where all the reaction rates are set to 1.0. With the inclusion of this network, we have obtained four different networks; three given by TOM, RI, and CLI, and a control network. For each one of these four networks, we have run 10 simulations. For each flux edge in a network, we computed the mean of 10 simulations, and then normalised these mean fluxes according to the maximum flux of each network.

Our simulations resulted in the normalised flux values listed in Table 6, where we have considered a spectrum of inhibition constants. Figure 2 provides a summary of the data in Table 6 with respect to the effect of varying inhibition constant from $p = 10^{-4}$ to $p = 1.0$. We observe that the inhibition constant p does not have a significant impact in general. More interestingly, the variations in p affect the versions of CRN that are instantiated with different measures similarly. Most of the fluxes are affected to an extent of 0.04% of maximum flux, and the greatest effect is on the fluxes that feed reaction 15 or compete with these fluxes, which however do not exceed 16% even with $p = 10^{-4}$, and these greater effects are pronounced at the lower end of the spectrum. Figure 3 displays the fluxes with $p = 10^{-2}$ for TOM, RI and CLI measures.

To compare the impact of the different measures on the simulations and the resulting fluxes, we have computed the distance between the results with different networks. We define this as the sum of squared distances between normalised fluxes. That is, given that $\mathcal{F}_1$ and $\mathcal{F}_2$ are flux graphs as in Fig. 3, for each flux edge from a reaction x to reaction y with w_1 in $\mathcal{F}_1$ and w_2 in $\mathcal{F}_2$, we compute the sum of the values $(w_1 - w_2)^2$. If a flux edge does not exist, its weight is 0.

$$\sum_{\substack{w_1 \in \mathcal{F}_1 \\ w_2 \in \mathcal{F}_2}} (w_1 - w_2)^2$$

For this network, we observe in Table 2 that the inhibition constant does not play a significant role in distinguishing the effect of different measures,

Table 2. The distances between flux graphs for the Dopamine network with the measures TOM, RI, CLI and control, which assigns the rate $r = 1.0$ to all the reactions.

p	TOM & 1.0	RI & 1.0	CLI & 1.0	TOM & RI	TOM & CLI	RI & CLI
10^{-4}	0.24	0.76	0.14	0.19	0.65	1.42
10^{-2}	0.24	0.78	0.15	0.2	0.66	1.47
1.0	0.24	0.8	0.14	0.21	0.64	1.46

Table 3. The ranking (#) of pathways delivered by measures RI, TOM, CLI and 1.0.

RI #	Pathway	RI flux	TOM #	CLI #	1.0 #
1	$0 \to 3 \to 19 \to 5 \to 8 \to 9 \to 10 \to 11$	0.83	1	1	1
2	$0 \to 1 \to 4 \to 5 \to 8 \to 9 \to 10 \to 11$	0.83	2	2	2
3	$0 \to 3 \to 19 \to 7 \to 17 \to 13 \to 14 \to 16$	0.74	3	6	5
4	$0 \to 1 \to 4 \to 7 \to 17 \to 13 \to 14 \to 16$	0.74	4	7	4
5	$0 \to 2 \to 18 \to 5 \to 8 \to 9 \to 10 \to 11$	0.73	5	5	3
6	$0 \to 3 \to 19 \to 7 \to 17 \to 13 \to 15$	0.69	6	3	8
7	$0 \to 1 \to 4 \to 7 \to 17 \to 13 \to 15$	0.69	7	4	7
8	$0 \to 2 \to 18 \to 7 \to 17 \to 13 \to 14 \to 16$	0.64	8	9	6
9	$0 \to 2 \to 18 \to 7 \to 17 \to 13 \to 15$	0.58	9	8	9
10	$0 \to 2 \to 15 \to$	0.45	10	10	10

which confirms our observations above. We observe that RI measure provides the greatest distinction from the control network and CLI provides the smallest distinction. The much larger distance between CLI and RI confirms this observation. Moreover, RI and TOM appear similar. Based on these observations, Table 3 enumerates the flux pathways by ranking them according to their mean fluxes, where RI measure is used as reference. As indicated by the observations in Table 2, all measures agree on the first two rankings and RI and TOM have the same rankings, which are different from those with the control network.

3   A Case Study: Escherichia coli Transcription Network

We have applied our method to the *Escherichia coli* transcription network version 10.5 reported in the RegulonDB [4] with the date 13 September 2018, which is depicted in Fig. 4. In this network, the distribution of the nodes with respect to their frequency in regulations follows a power-law, whereby 1610 of the 1886 proteins participate in not more than 5 regulations. As listed in Table 7, CRP has the highest frequency as it participates in 585 regulations, followed by FNR with 322, IHF with 259 and H-NS with 195 regulations.

We applied the four measures given by TOM, RI and CLI as well as the control model with all the rates set to 1.0, and considered the inhibition constants

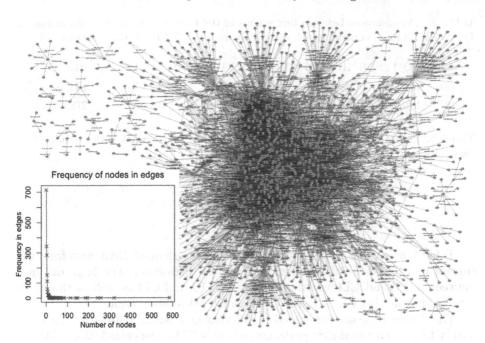

Fig. 4. Escherichia coli transcription network as reported in [4], rendered by Cytoscape [19]. The network consists of 1886 nodes, which are regulated by 4639 edges. The 2338 activation edges are denoted by green, whereas the 2301 repression edges are denoted by red. 207 of the nodes are transcription factors and 63 are at the root position. The graph in the corner displays the frequency of nodes in the edges. (Color figure online)

$p = 10^{-2}$ and $p = 1.0$, and this way obtained 8 different versions of the model. Each of these networks consist of 4639 reactions with 1886 species. We focused our investigation on the pathways initiated by CRP as this transcription factor has the highest frequency among all the 63. We performed 10 simulations with an initial value of 10000 CRP molecules for all of the 8 cases. Each of these simulations resulted in 1000 to 1500 flux edges. For each case, we took the mean of each flux edge given by the 10 simulations. We then normalised each one of the 8 flux graphs with the maximum flux in that graph.

For a comparison, we first computed the squared distance between the 8 flux graphs. To emphasise the effect of different measures, we have taken the sum of the fluxes for each species in flux graphs and computed the squared distance on these sums. The differences between TOM, RI, CLI and 1.0 model for each of the $p = 10^{-2}$ and $p = 1.0$ values are listed in Table 4. The differences between $p = 10^{-2}$ and $p = 1.0$ for each of TOM, RI and CLI and 1.0 are listed in Table 5. We observe in Tables 4 and 5 that the inhibition constant does not play a significant role in distinguishing the effect of different measures as before with the exception of 1.0 network. This observation confirms that the rates provided by the measures plays a more significant role in determining the fluxes in comparison to the inhibition constant. However, in the control model, the inhibition constant plays a greater role in determining the system behaviour.

Table 4. The distances between flux graphs of the E. coli network with the measures TOM, RI, CLI and control, which assigns the rate $r = 1.0$ to all the reactions.

p	TOM & 1.0	RI & 1.0	CLI & 1.0	TOM & RI	TOM & CLI	RI & CLI
1.0	17620	21417	91637	191	28897	24470
10^{-2}	19579	24360	104359	267	33543	27897

Table 5. The distances between flux graphs of the E. coli network with the inhibition constants $p = 1.0$ and $p = 10^{-2}$.

p	TOM	RI	CLI	1.0
1.0 & 10^{-2}	282	204	13	106662

Table 4 indicates that CLI measure provides the greatest distinction from the control model and TOM provides the smallest distinction. The large distance between TOM and CLI and the one between RI and CLI as well as the much smaller distance between RI and TOM confirm this observation.

The different measures resulted in different numbers of pathways. With $p = 0.01$, CLI has generated 1388 pathways, whereas RI has generated 998, TOM has generated 1340 and the control network has generated 1535. With $p = 1.0$, CLI has generated 1285 pathways, whereas RI has generated 1210 pathways, TOM has generated 1251, and the control network has generated 1535 pathways. The resulting list of pathways for all the 8 cases can be downloaded together with all the scripts that are used to apply the methods above.[2]

4   Discussion

We have proposed a method for enumerating dominant pathways in biological networks that can be represented as directed graphs consisting of activation and repression edges. Our analysis combines methods from static graph theoretical considerations in the literature with those for dynamical systems, based on simulations. Our method emphasises the inherent stochasticity in biological processes as well as the notion that cellular signals are relayed predominantly through highly connected nodes and pathways of least biochemical resistance.

The stochastic simulations in our examples result in individual simulation trajectories that expose the noise in the system. The notion of stochastic flux, delivered by these simulations, provides a direct quantification of information flow, for any time interval, including the transient states. However, averaging over many simulations as in the examples above dampens the stochastic noise. If a deterministic notion of information flow can be characterised, linear noise approximation simulations [2] or deterministic ODE simulations can be considered for the analysis of the systems where the stochastic noise is less of a concern.

---

[2] ozan-k.com/pathways.zip.

Table 6. Mean normalised fluxes obtained from 10 simulations for each CRN in Table 1 instantiated with rates given by TOM, RI, CLI and 1.0, and with inhibition constants of $p = 10^{-4}$, $p = 10^{-2}$ and 1.0. The table thus summarises 120 simulations.

	TOM			RI			CLI			1.0		
Fluxes /p	10^{-4}	10^{-2}	1.0	10^{-4}	10^{-2}	1.0	10^{-4}	10^{-2}	1.0	10^{-4}	10^{-2}	1.0
$0 \to 1$	0.78	0.82	0.82	0.76	0.8	0.8	0.79	0.83	0.83	0.69	0.73	0.73
$0 \to 2$	0.52	0.55	0.55	0.62	0.66	0.66	0.69	0.72	0.72	0.69	0.73	0.73
$0 \to 3$	0.78	0.82	0.83	0.76	0.8	0.8	0.79	0.83	0.83	0.69	0.73	0.73
$1 \to 4$	0.78	0.82	0.82	0.76	0.8	0.8	0.78	0.82	0.82	0.69	0.73	0.73
$2 \to 15$	0.07	0.19	0.2	0.11	0.23	0.23	0.11	0.23	0.24	0.08	0.19	0.2
$2 \to 18$	0.44	0.36	0.35	0.51	0.43	0.42	0.58	0.49	0.48	0.61	0.54	0.54
$3 \to 19$	0.78	0.82	0.83	0.76	0.8	0.8	0.78	0.82	0.82	0.69	0.73	0.73
$4 \to 5$	0.39	0.41	0.41	0.38	0.4	0.4	0.38	0.4	0.4	0.35	0.37	0.37
$4 \to 7$	0.39	0.41	0.41	0.38	0.4	0.4	0.38	0.4	0.4	0.35	0.36	0.37
$5 \to 8$	1.0	1.0	1.0	1.0	1.0	1.0	1.0	1.0	1.0	1.0	1.0	1.0
$7 \to 17$	1.0	1.0	1.0	1.0	0.99	1.0	0.98	0.99	0.99	1.0	1.0	1.0
$8 \to 9$	0.99	0.99	1.0	0.98	0.98	0.98	0.91	0.91	0.91	1.0	1.0	1.0
$9 \to 10$	0.98	0.98	0.98	0.94	0.95	0.95	0.77	0.78	0.78	1.0	1.0	1.0
$10 \to 11$	0.95	0.95	0.95	0.88	0.89	0.89	0.61	0.62	0.62	1.0	1.0	1.0
$13 \to 14$	0.88	0.77	0.76	0.77	0.65	0.65	0.57	0.45	0.45	0.92	0.81	0.8
$13 \to 15$	0.07	0.19	0.2	0.11	0.23	0.23	0.11	0.23	0.24	0.08	0.19	0.2
$14 \to 16$	0.83	0.72	0.71	0.69	0.57	0.57	0.42	0.31	0.3	0.92	0.81	0.8
$17 \to 13$	0.98	0.98	0.98	0.94	0.94	0.95	0.84	0.84	0.85	1.0	1.0	1.0
$18 \to 5$	0.22	0.18	0.17	0.26	0.21	0.21	0.29	0.24	0.24	0.31	0.27	0.27
$18 \to 7$	0.22	0.18	0.17	0.26	0.21	0.21	0.29	0.24	0.24	0.31	0.27	0.27
$19 \to 5$	0.39	0.41	0.41	0.38	0.4	0.4	0.38	0.4	0.4	0.35	0.36	0.37
$19 \to 7$	0.39	0.41	0.41	0.38	0.4	0.4	0.38	0.4	0.4	0.35	0.37	0.37

As evidenced by our case study on *E. coli* network, the measures, TOM, RI, and CLI have a significant effect on determining the dominant pathways. In this regard, a more extensive evaluation of these measures as well as others in the literature is a topic of further investigation. Moreover, the ranking of the pathways is subject to parameters such as pathway length and flux strengths at various segments, which can change the ranking. An evaluation of these parameters in the context of biological evidence for the *E. coli* network and in applications to other large networks are topics of future work.

Table 7. The frequency of transcription factors (TF) in the E. coli network.

Freq.	TF	Freq.	TF	Freq.	TF	Freq.	TF	Freq.	TF	Freq.	TF
585	CRP	259	IHF	84	NsrR	66	NarP	46	ModE	16	RcsAB
322	FNR	195	H-NS	82	FlhDC	56	NtrC	34	SlyA	16	rcsB-BglJ
13	HypT,	12	GntR	11	NanR	9	HU	9	AllR	8	MatA
13	BasR	12	HprR	9	NrdR	9	RcdA	9	GadE-RcsB	8	PgrR
8	DicA	7	UlaR	7	DeoR	6	GatR	6	Zur	5	AscG
7	TdcR	7	CueR	6	GlrR	6	SdiA	5	CsiR	5	MntR
5	HipAB	4	BluR	4	KdpE	4	NadR	3	DinJ-YafQ	3	MazE-MazF
5	BirA	4	McbR	4	AtoC	3	RclR	3	relB-RelE	2	RqhC
2	FabR	2	EnvR	2	KdgR	2	RtcR	2	XapR	1	YpdB
2	EnvY	2	EbgR	2	BCCP	2	HigB-HigA	2	YefM-YoeB	1	ZntR
1	UhpA	1	BtsR	1	YhaJ						

References

1. Albert, R.: Scale-free networks in cell biology. J. Cell Sci. **118**, 4947–4957 (2005)
2. Cardelli, L., Kwiatkowska, M., Laurenti, L.: Stochastic analysis of chemical reaction networks using linear noise approximation. Biosystems **149**, 26–33 (2016)
3. Erhard, F., Friedel, C.C., Zimmer, R.: FERN: a Java framework for stochastic simulation and evaluation of reaction networks. BMC Bioinform. **9**, 356 (2008)
4. Gama-Castro, S., et al.: RegulonDB version 9.0: high-level integration of gene regulation, coexpression, motif clustering and beyond. Nucleic Acids Res. **44**(D1), D133–D143 (2016)
5. Gillespie, D.T.: Exact stochastic simulation of coupled chemical reactions. J. Phys. Chem. **81**(25), 2340–2361 (1977)
6. Gillespie, D.T.: Approximate accelerated stochastic simulation of chemically reacting systems. J. Chem. Phys. **115**(4), 1716 (2001)
7. Kahramanoğulları, O.: On linear logic planning and concurrency. Inf. Comput. **207**, 1229–1258 (2009)
8. Kahramanoğulları, O.: Quantifying information flow in chemical reaction networks. In: Figueiredo, D., Martín-Vide, C., Pratas, D., Vega-Rodríguez, M.A. (eds.) AlCoB 2017. LNCS, vol. 10252, pp. 155–166. Springer, Cham (2017). https://doi.org/10.1007/978-3-319-58163-7_11
9. Kahramanoğulları, O., Lynch, J.: Stochastic flux analysis of chemical reaction networks. BMC Syst. Biol. **7**, 133 (2013)
10. Khatri, P., Sirota, M., Butte, A.J.: Ten years of pathway analysis: current approaches and outstanding challenges. PLoS Comput. Biol. **8**(2), e1002375 (2012)
11. Kuwahara, H., Mura, I.: An efficient and exact stochastic simulation method to analyze rare events in biochemical systems. J. Chem. Phys. **129**(16), 10B619 (2008)
12. Ma, S., Jiang, T., Jiang, R.: Differential regulation enrichment analysis via the integration of transcriptional regulatory network and gene expression data. Bioinformatics **31**(4), 563–571 (2015)
13. Ma'ayan, A., et al.: Formation of regulatory patterns during signal propagation in a Mammalian cellular network. Science **309**(5737), 1078–83 (2005)

14. Nielsen, M., Plotkin, G., Winskel, G.: Event structures and domains, part 1. Theor. Comput. Sci. **5**(3), 223–256 (1981)
15. Persson, O.: Identifying research themes with weighted direct citation links. J. Informetr. **4**(3), 415–422 (2010)
16. Planes, F.J., Beasley, J.E.: A critical examination of stoichiometric and path-finding approaches to metabolic pathways. Brief. Bioinform. **9**(5), 422–436 (2008)
17. Randić, M.: Characterization of molecular branching. J. Am. Chem. Soc. **97**(23), 6609–6615 (1975)
18. Ravasz, E., et al.: Hierarchical organization of modularity in metabolic networks. Science **297**, 1551–1555 (2002)
19. Shannon, P., et al.: Cytoscape: a software environment for integrated models of biomolecular interaction networks. Genome Res. **13**(11), 2498–504 (2003)
20. Zubarev, R.A., et al.: Identification of dominant signaling pathways from proteomics expression data. J. Proteomics **71**(1), 89–96 (2008)

Comparing Different Graphlet Measures
for Evaluating Network Model Fits
to BioGRID PPI Networks

Sridevi Maharaj$^{(\boxtimes)}$ ⓘ, Zarin Ohiba ⓘ, and Wayne Hayes ⓘ

Department of Computer Science, University of California, Irvine, Irvine, USA
{sridevi.m,zohiba,whayes}@uci.edu

Abstract. The network structure of protein-protein interaction (PPI) networks has been studied for over a decade. Many theoretical models have been proposed to model PPI networks, but continuing noise and incompleteness in these networks make conclusions difficult. Graphlet-based measures are believed to be among the strongest, most discerning and sensitive network comparison tools available. Several graphlet-based measures have been proposed to measure topological agreement between networks and models, with little work done to compare the measures themselves. The last modeling attempt was 4 years ago; it is time for an update. Using Sept. 2018 BioGRID, we fit eight theoretical models to nine BioGRID networks using four different graphlet-based measures. We find the following: (1) Graph Kernel is the best measure based on ROC and AUPR curves; (2) most graphlet measures disagree on the ordering of the data-model fits, although most agree on the top two (STICKY and Hyperbolic Geometric) and bottom two (ER and GEO) models, in direct contradiction to the 4-years-ago conclusion that GEO models are best; (3) the STICKY model is overall the best fit for these PPI networks but the Hyperbolic Geometric model is a better fit than STICKY on 4 species; and (4) even the best models provide p-values for BioGRID that are many orders of magnitude smaller than 1, thus failing any reasonable hypothesis test. We conclude that in spite of STICKY being the best fit, all BioGRID networks fail all hypothesis tests against all existing models, using all existing graphlet-based measures. Further work is needed to discover whether the data or the models are at fault.

Keywords: Protein interaction · Biological networks · Systems biology

1   Introduction

Networks have been used for decades to model biological processes and interactions such as transcription [40] and gene regulatory networks [14], proteome-scale interactions in human cells [20], and brain connectomes [24]. A protein-protein interaction (PPI) network is a graph whose nodes are proteins and edges represent observed physical interactions between the proteins (nodes). Many network

ⓒ Springer Nature Switzerland AG 2019
I. Holmes et al. (Eds.): AlCoB 2019, LNBI 11488, pp. 52–67, 2019.
https://doi.org/10.1007/978-3-030-18174-1_4

models have been suggested to better comprehend and describe the connection patterns within PPI networks. Early work suggested that the degree distribution followed a scale-free law [3,39]. While evolution certainly must play a role in structuring biological networks [4,46], different modules may show significantly different structures [21,30]. In this paper, we follow up and update previous studies [9,13,33], primarily using *graphlet* measurements. Although graphlets have been used previously to model PPIs, the amount of data available has increased dramatically in recent years, and no comprehensive evaluation of the measures themselves has been done. These together justify a revisit to the question of which models best fit the current data.

1.1 Graphlets

Given a graph $G(n, m)$ on n nodes and m edges, a *k-graphlet* is an induced subgraph g on any set of k connected nodes from G, where k is typically between 2–5 [32]. Figure 1 shows all the graphlets on 2, 3, 4, and 5 nodes including their *automorphism orbits* [37]. Automorphisms enumerate all the different ways of drawing the same graph. For example, we can draw G_{15} as it is drawn in Fig. 1 or as a star, but both drawings are cycles of length 5. The graphlet-orbit *signature* of a node v is a vector of counts of each graphlet orbit to which v belongs [26].

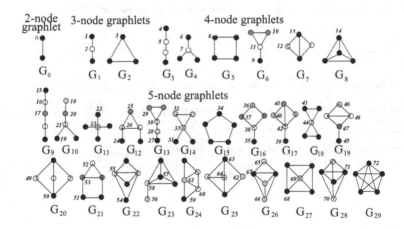

Fig. 1. All 2, 3, 4, and 5-node graphlets. The numbers in normal font represent the graphlet ordering of [32]. Within each graphlet, *automorphism orbits*—nodes that are topologically identical within the graphlet—have the same shade [37]. There are 73 distinct orbits, numbered 0 through 72, identified by italicized numbers next to one of the nodes of each orbit.

Graphlets have become very popular as a way of quantifying local structure within biological networks. Graphlets and their orbits have been used to (i) aid global alignments [18,19,22,23,27,44], (ii) perform alignment-free comparison of networks [9,15,32,36,48] and, (iii) recover both functional and phylogenetic information [6,19].

1.2   Graphlet-Based Network Comparison Measures

Graphlet-based network comparison measures can be categorized as distance measures (smaller is better), and similarity measures (larger is better). Distance measures include: Relative Graphlet Frequency Distance (RGFD), which compares log counts of graphlets within two networks [32]; and Graphlet Correlation Distance (GCD), which is the Euclidean distance between correlations of pairs of graphlet orbits [47]. Similarity (or agreement) measures include: Graphlet Degree Distribution Agreement (GDDA), which compares distributions of each orbit within two networks [37]; and Graphlet Kernel (GK), which is the dot product of normalized graphlet counts [41]. Complete formulae for RGFD, GCD, GDDA and, GK are shown in Table 1.

Table 1. The formulae for computing RGFD, GCD, GK, and GDDA. The top two measures are distance measures, so smaller is better; while the bottom two measures are similarities, so larger is better. Consider networks G and H. For RGFD, the total number of each type of graphlet for each network is counted and stored in vectors u and v, which are used in the computation. For GCD, pairwise orbit correlations across nodes are computed. Computing GDDA begins with finding the distribution of orbits of type j, $D_G^j(k)$ in each graph: count the number of nodes which belong to orbits of type j exactly k times, for $k \in \mathbb{Z}_{\geq 0}$ and divide each $D_G^j(k)$ by k, then normalize and then finally, the norm. For GK, we use the same vectors as for RGFD.

Measure	Input variables	Formula
Relative Graphlet Frequency Distance (RGFD)	Graphlet counts $\mathbf{u}$, $\mathbf{v}$	$\sum_i \| -log(\frac{u_i}{\sum_j u_j}) + log(\frac{v_i}{\sum_j v_j})\|$
Graphlet Correlation Distance (GCD)	Correlation matrices of signatures, C_1, C_2	$\sum_{i,j}(C_1[i,j] - C_2[i,j])^2$
Graphlet Degree Distribution Agreement (GDDA)	Graphlet orbit degree distributions D_G^j, D_H^j $S_G^j(k) = \frac{D_G^j(k)}{k}$, $N_G^j(k) = \frac{S_G^j(k)}{\sum_k S_G^j(k)}$	$1 - \sqrt{(\sum_{k=0}^{\infty}(N_G^j(k) - N_H^j(k))^2)}$
Graphlet Kernel (GK)	Graphlet counts $\mathbf{u}$, $\mathbf{v}$ or unit vectors of the graphlet counts	$\frac{u \cdot v}{\|\|u\|\| \cdot \|\|v\|\|}$

1.3   Models

Most network models are created with a particular application in mind. The 8 network models—ER, ERDD, GEO, GEOGD, SF, SFGD, STICKY, and HGG— used in this study are described in Table 2. The original ER paper proves many theorems about the graph model but makes no reference to a real physical system [8]. The primary strength of the ER model remains that theorems are often more easily proved in ER graphs than otherwise. GEO graphs can be used to describe any system in which physical proximity plays a role, for example to describe contact points in folded bio-molecules [45]. The phenomenon of self-similarity or *scale-free* (SF) nature has been observed to describe many real systems, including the distribution of asteroid sizes [29] and time between extinction events [42].

Table 2. Models used in this study are presented below.

Model name	Abbrev	Description
Erdös-Rényi	ER	Every pair of nodes (u, v) are connected with probability p [8]
Erdös-Rényi Degree Distribution	ERDD	Nodes assigned stub counts according to input degree distribution Edge placed between randomly selected nodes with non-zero stub count [25]
Scale Free	SF	Degree distribution asympotically follows power law [2]
Scale Free Gene Duplication and Divergence	SFGD	New node is attached to parent node with probability p New node is attached to parent's neighbors with probability q [43]
Geometric	GEO	Nodes are points in a metric space Nodes are connected if within radius ϵ of each other [28]
Geometric with Gene Duplication and Mutation	GEOGD	Nodes are points in a metric space Nodes further than ϵ radius from each other may be connected [34]
STICKY	STICKY	Nodes assigned stickiness index based on relative degree according to input degree distribution. Product of stickiness indices between nodes determine their probability of interaction [33]
Hyperbolic Geometric	HGG	Nodes are points on a hyperbolic disk. Connection probability between two nodes is a function of their distance in hyperbolic space [17]

HGG models graphs which capture some notion of metric structure and have a heterogeneous degree distribution [17]. To better model real systems, sometimes the above models were re-created under further constraints by adding gene duplication (GD). The SFGD and GEOGD models were designed as null models for PPI networks [34,43]. The STICKY model starts with the observation that proteins have binding domains, and the number of binding domains increases with the size of the protein [33]. Thus larger proteins have more binding domains—they are more "sticky"—and model proteins are deemed to interact with a probability proportional to the product of their stickiness indices.

1.4   Comparing Models to Data

Early work used the degree and k-hop distributions to argue that PPI networks were SFGD [43]. Graphlet-based measures later showed using RGFD [32,33] and GDDA [9,10,12,13,31,34,35] that yeast, fly, worm, and human were better fit by the following models in order of goodness (best to worst): STICKY [9,12, 13,31,33], GEOGD [34], GEO [10,31,35], SFGD, SF, ERDD, ER [34]. Note in particular that adding gene duplication to GEO and SF made both fit better [9,31] though still not as well as STICKY [9]. In terms of measures, all graphlet-based modeling has been tested using only RGFD and GDDA, while GK [41] and GCD [47] were developed later and may be better. In addition, the HGG model has never been tested against PPI networks.

In summary, it has been 4 years since the last PPI modeling study has been done. During that time, the amount of network data has exploded; new measures and models have been introduced; and no comprehensive study has been done across all species, all available models, and all available measures. In this paper we aim to rectify these deficiencies.

Table 3. BioGRID PPI networks [5] used in this study, ordered by edge density, downloaded in Sept. 2018.

Species (*Latin name*)	Code	Common name	Num nodes n	Num edges m	Density	Mean degree
Saccharomyces cerevisiae	SC	Brewer's Yeast	6879	104719	0.00443	30.44
Schizosaccharomyces pombe	SP	Fission Yeast	2951	8754	0.00201	6.09
Drosophila melanogaster	DM	Fruit fly	8836	46288	0.00119	10.48
Homo sapiens	HS	Human	22376	277940	0.00111	24.85
Caenorhabditis elegans	CE	Roundworm	3276	5638	0.00105	3.46
Arabidopsis thaliana	AT	Thale Cress	9571	35253	0.00077	7.37
Rattus norvegicus	RN	House Rat	3569	4952	0.00046	2.80
Mus musculus	MM	House Mouse	12817	37915	0.00046	5.91
Escherichia coli	EC	Bacteria	2044	12800	0.000025	12.52

1.5   Our Contribution

In this study, we use the 4 measures of Table 1 to re-evaluate the 8 models of Table 2 on the newest data for the 9 largest species shown in Table 3 available from BioGRID [5] downloaded September 2018, for 9 species. We reconfirm that STICKY is still (usually) the *overall* best-fitting model for all 9 networks, but the HGG model is a better fit than STICKY on 4 species, and SF is a good second fit on other species. Surprisingly, we find that the GEO model, which was previously found to be almost on par with the STICKY model [9,13,31,35], is now roughly tied with ER as the worst-performing model. We further evaluate the network comparison measures of Table 1, and find that (1) they differ substantially in their evaluations of model-data fits, and that (2) the Graphlet Kernel significantly outperforms all other measures at classifying models based on the area under a precision-recall curve (AUPR). The need for such a comparison is two-fold: first, such comparisons have not previously included the Graphlet

Kernel; second, most previous graphlet-based studies utilized only RGFD and GDDA measures to determine null models for biological networks, but not GK or GCD. Hence, we provide a far more comprehensive analysis than previous studies.

2   Methods

We created 500 synthetic versions of each of the 9 BioGRID PPI networks (see Table 3) using each of the 8 models listed in Table 2. To generate ER, SF and GEO networks, we set the relevant parameters (such as number of nodes, density, radius, attachment index, etc) so that the resulting graphs matched the size and density of each of the PPI networks. To generate HGG networks, we used the implementation of [1] and set the number of nodes, average degree and expected power-law exponent to match those of the PPI networks. We generated 9 sets of SFGD networks for different values of p (see Table 2) ranging from 0.1 to 0.9 in increments of 0.1. Following [34], for each of these sets, we exactly matched the number of nodes of the original PPI network, and did a binary search on the corresponding q value (cf. Sect. 1.3) until the synthetic graph contained within 1% of the number of edges in the real network. We generated GEOGD synthetic networks using both the expansion and probability cutoff methods described in [34], incrementing the probability by 0.1 from 0.1 to 0.9. We thus created a total of 112,500 synthetic networks.

We ran ORCA [11] on all the above networks to count all of the graphlets of size $k = 2, 3, 4$, and 5, and converted its graphlet orbit signature matrix output to graphlet counts. We computed RGFD, GCD, GDDA, and GK measures between the synthetic model networks and the original BioGRID networks. This enabled us to observe which models fit best, and to compare the measures to each other.

3   Results

3.1   Assessing Quality of Each Measure

Before drawing conclusions from comparison with the PPI networks, we assess how well each network comparison measure is able to distinguish between networks of different types. To do this, we employ standard Area Under the Precision-Recall (AUPR) curves and Receiver Operating Characteristic (ROC) curves. For various values of a threshold ϵ, if the distance (similarity) between two networks is smaller (greater) than ϵ, we categorize those two networks together. The area under these curves measure the quality of grouping according to the network comparison measure used. Figure 2 shows that the GK measure has the best AUPR and AUROC overall: (i) GK outperforms non-graphlet based measures, thereby corroborating previous studies that graphlet topology is important in the study of network comparison [9], (ii) GK also outperforms other graphlet-based measures, suggesting that amongst these measures, it is the one most suitable for distinguishing between networks.

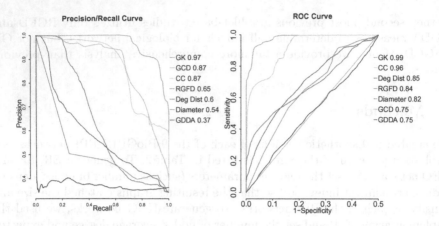

Fig. 2. We computed the pairwise distances and similarities between 1500 randomly selected networks from our synthetic network dataset, 300 each from 5 classes (ER, ERDD, GEO, STICKY and, SF), using the graphlet-based measures of Table 1 and some non-graphlet-based measures, namely clustering coefficient (CC), degree distribution (Deg Dist) and diameter. Then, we thresholded the distances (similarities) scores in 0.1% increments of the range of pairwise distances (similarities), and computed the AUPR and AUROC for each network comparison measure. Precision is defined to be $\frac{TP}{TP+FP}$, recall or sensitivity is $\frac{TP}{TP+FN}$, and specificity is $\frac{TN}{TN+FP}$, where $TP =$ true positives, $FP =$ false positives, $FN =$ false negatives and, $TN =$ true negatives. The AUPR curve may be more indicative of measure ordering because PR is known to be more appropriate than ROC when there are many negative cases (in our case, there are many more pairs of networks not belonging to the same class than there are pairs in the same class of networks) [7]. Overall, the GK measure has by far the greatest AUPR, surpassing that of GCD, while GDDA has the lowest. GK also has the greater AUROC while GDDA and GCD have the lowest.

3.2 Assessment of Fits to PPIs by Graphlet-Based Measures

Figure 3 depicts the RGFD, GCD, GDDA, and GK scores between the model networks and real network of each species. According to the GK measure, HGG and STICKY are by far the best fitting models but according to the RGFD measure, the graphlet topology of the 9 networks is fit best by STICKY and then by the SF and SFGD models. HGG performs mediocre on most measures but exhibits high similarity with the PPI networks when assessed by the GK measure. Using the GDDA and GCD measures, the best overall fits after STICKY are ERDD, SF, SFGD and HGG. All measures unanimously assess GEO, GEOGD and ER to be the worst fits. Older versions of all of these networks were already modeled well by STICKY graphs and we make the same observation here. This demonstrates further that the STICKY model is a plausible model for PPI networks even as the data evolve. We find that under GDDA, the order of suitability in modeling the structure of the species CE, DM, HS and SC networks has changed since the study of [9]. The four species were modeled best by STICKY, SFGD and GEOGD (in that order) but using the current data, in order of best fit,

CE is best modeled by STICKY, ERDD and then SFGD; DM is best modeled by STICKY, ERDD and then SF; HS is best modeled by STICKY, SFGD and then SF; and SC by STICKY, SF and then SFGD. Both GDDA and GCD use graphlet orbital information, and generally agree on the top three best-suited models but often disagree on the ordering (agreeing in order only on HS).

Since the most recent similar study to ours is [13], it is worth performing a direct comparison. Those authors concentrated on how the network structure changes according to interaction-screening technology; they studied only one species—yeast, because that was the most complete PPI network at the time—and used only GDDA as the comparison measure. One of their data sets was "BioGRID Yeast"—their Fig. 1, "biogrid" column. Direct comparison of their biogrid column to our SC column, using the GDDA sub-figure of our Fig. 3, we see that there is actually broad agreement: STICKY is by far the best, ER is by far worst, and GEO falls roughly between the two. All that has changed between their study and ours, in this one column of overlap, is the ordering of the other models. Where we disagree with [13] is in which measure is best to use: our Fig. 2 clearly demonstrates that GDDA is not only a bad measure for judging model fit, it is the *worst* measure among all the graphlet-based measures for this purpose.

Since Fig. 2 suggests GK is the best measure, we focus on its results in Table 4. The high scores on most networks together with the overall best-fit results (STICKY, HGG, SF, SFGD) may be suggesting that the PPI network structures are exhibiting patterns of more than one network model. Note that, we do not claim there is one model that fits all, but that STICKY is the best fit for 5 (DM, EC, HS, MM, RN) of the 9 PPI networks used in this study, HGG is the best fit for 4 (AT, CE, SC, SP) of the 9 PPI species, overall SF is also a good fit for all models except SC and SFGD is a good fit for 3 species (CE, DM, EC).

3.3   Mixed Agreement Across Measures

One striking feature of all these network measures is that their behavioural trends do not match each other, even though they all use the same graphlet or orbit counts to quantify the similarity (or difference) in networks. For example, the best fitting models under GDDA and GCD do not agree with GK and RGFD, and the four measures often provide wildly different orderings of the models from best-to-worst (other than putting STICKY on top and GEO on the bottom). Table 5 attempts to quantify the relationships between these network comparison measures with the Pearson correlation coefficients over all models and all species for each pair of measures. Some of the disagreements between different measures may be due to how they treat individual graphlets: the RGFD and GK measures may be influenced most by the graphlets with the highest counts, while graphlets with relatively lower counts have a smaller weight on the overall score. However, small differences in orbit counts at each node may cause a bigger GCD difference because orbit trends are compared with every other orbit. In addition, [9] previously argued that GDDA is sensitive. Hence, GCD and GDDA may respond more sensitively to graphlet (orbit) count differences than the other measures.

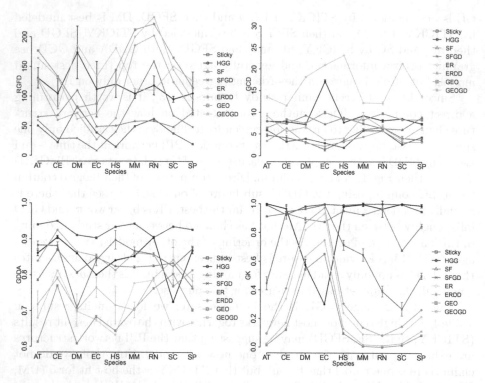

Fig. 3. The average computed score comparing species-vs-model networks for the distance measures RGFD and GCD (top), and the two similarity measures GDDA and GK (bottom). Error bars along each curve depict 1σ standard deviations of the species-vs-model scores across the 500 synthetic networks for that (species, model) pair. The standard deviations are larger for HGG, SFGD and GEOGD possibly because there are more random choices involved in matching the input parameters such as average degree and number of edges. For each species, within the 10 GEOGD and 9 SFGD types, the differences were very small (on order of magnitude 10^{-1}). Hence, we have plotted only the best performing GEOGD and SFGD models from among the GEOGD expansion and GEOGD probabilistic (p ranging from 0.1 to 0.9) networks and the SFGD models (p ranging from 0.1 to 0.9), respectively. Note that all measures appear to agree that STICKY is one of the best models, often being among the lowest curve in the difference measures, and the highest curve in the similarity measures. HGG is on par with STICKY on most species according to the GK measure. Interestingly, the curves in the GK plot cross each other far less frequently than with the other measures, suggesting that GK maintains monotonicity in model quality, across all species, much better than the other measures; this property re-enforces, along with the PR and ROC curves, that GK gives a clean and precise measure of model quality.

Since STICKY is the overall best fit of the PPI networks, we further examine its fit to the BioGRID networks. Following the non-parametric test of [38] (see also [9]), for each species, and for each graphlet type, we computed the mean and standard deviation of the graphlet count across all 500 synthetic STICKY

Table 4. On the assumption that GK is the best measure as depicted in Fig. 2, we show its similarity scores below (score of 1 is best). The rows are ordered best-to-worst according to the "Average" column. Within each column, the best value is boldfaced and the second-best value is italicized, with up to two if there are close ties. The GK scores indicate that STICKY is the best model for most PPI networks, followed by SF. While HGG has the 3rd overall average GK score, it performs as well as STICKY on HS, MM, and RN, and is a better match for AT, CE, SC and SP than the STICKY model. The SFGD model is a good fit on some PPI networks. Contrary to previous findings, GEOGD and GEO models are not good fits for *any* of these 9 PPI networks.

Model	Species									
	AT	CE	DM	EC	HS	MM	RN	SC	SP	Average
STICKY	*0.991*	*0.923*	**0.981**	**0.996**	**0.990**	**0.998**	**0.999**	*0.995*	*0.948*	**0.980**
SF	*0.912*	*0.931*	0.896	0.873	*0.950*	*0.924*	*0.968*	0.677	0.844	*0.886*
HGG	**0.999**	**0.966**	0.575	0.300	*0.978*	*0.995*	0.965	**0.999**	**0.987**	*0.863*
SFGD	0.441	*0.953*	*0.964*	*0.993*	0.717	0.650	0.388	0.262	0.500	0.652
ERDD	0.093	0.411	0.819	0.973	0.305	0.138	0.082	0.146	0.264	0.359
ER	0.098	0.339	0.869	0.933	0.240	0.091	0.087	0.096	0.215	0.330
GEOGD	0.024	0.169	0.633	0.747	0.129	0.018	0.011	0.026	0.113	0.208
GEO	0.014	0.122	0.543	0.651	0.097	0.007	0.001	0.014	0.084	0.170

Table 5. How the various graphlet measures agree with each other, measured by the Pearson correlation ρ of their model-vs-data similarity (or difference) scores. The magnitude of the correlation depicts "amount" of agreement; the sign is relevant only in that it depicts whether the measures agree in direction (both similarities or both differences), or are opposite types (difference vs. similarity). The p-value is the probability that the observed correlation is due to chance. Thus, RGFD and GK have the greatest correlation, while GK and GCD have such a low correlation (-0.21) that the p-value—just 7.1%—means their agreement is barely distinguishable from random.

Measures	ρ	p-value
RGFD and GK	-0.668	1.5×10^{-10}
GCD and GDDA	-0.565	2.4×10^{-6}
GK and GDDA	0.5	7.5×10^{-6}
RGFD and GDDA	-0.482	1.8×10^{-5}
RGFD and GCD	0.460	4.7×10^{-5}
GK and GCD	-0.21	7.1×10^{-2}

networks. BioGRID's graphlet count is shown in Table 6 as a multiple of the standard deviation away from the mean count in the synthetics. This factor is small for graphlet 0 (the number of edges in the graph). However, the factors grow very large on some graphlets in each species (more than $10^5\sigma$ as is the case for DM), as shown in Table 6. This means that though STICKY is the *best* amongst the models examined in this study, it is still far from an *ideal* match to the current BioGRID networks.

Table 6. The mean and standard deviation of each graphlet's count in the STICKY synthetic networks, together with the deviation of the true count from the synthetic mean. For each species, the best-fitting and worst-fitting graphlets are shown; the absolute worst offenders (thousands of standard deviations away) are highlighted in bold. Though STICKY is generally the best structural match among all 7 models explored, the last column demonstrates that it is still far from an ideal fit to the current BioGRID networks. Note that graphlet G_{20} (cf. Fig. 1) is often the worst offender, although the significance of this observation is unclear.

Species	Graphlets (best,worst)	Mean STICKY count	σ of STICKY count	BioGRID count	BioGRID distance to STICKY mean
CE	0	5619	74	5638	0.26σ
	20	5572	886	21554	18σ
SP	27	33810	3057	32556	0.41σ
	20	62298	5221	330540	51.4σ
SC	6	6.18×10^8	6.09×10^6	618385434	0.06σ
	20	5.20×10^7	940839	1833805726	**1900σ**
EC	0	12784	110	12800	0.15σ
	29	14.6	8.85	2727	306σ
RN	25	3188	334	2945	0.73σ
	20	1405.2	218.206	53753	240σ
AT	21	2.87×10^7	912783	29801214	1.15σ
	20	1.18×10^6	54516	256721361	**4700σ**
DM	0	46268	206	46288	0.09σ
	29	4.29	3.79	1278120	**$337,000\sigma$**
HS	0	277199	514	277940	1.44σ
	20	1.86×10^8	2.75×10^6	1369437894	430σ
MM	0	37650	188	38075	2.26σ
	20	730868	34637	9773813	261σ

4   Fit of PPI Networks from the 2018 Update of the IID Database

We also present fits to 9 PPI networks from the Integrated Interactions Database (IID) [16], including 7 mammals which have not been studied before in the context of modeling PPI networks (see Table 7). While all the BioGRID networks described previously have edges derived solely from wet-lab experiments, it is known that the edges they contain are both highly biased, and far from complete [20]. The IID is a recent attempt to ameliorate these issues. IID networks contain edges not just from BioGRID, but from all experimental efforts to date; and in addition, edges predicted from sophisticated machine learning techniques. Thus, although partly machine-generated, the IID networks are currently a best-effort guess as to the approximate size and structure of true PPI networks. Thus, it is appropriate to analyze them using our techniques.

Due to the large size of the IID networks, the time complexity of generating HGG networks as well as other time constraints, we show the fits of these 9 species to STICKY, SF, GEO, ER and ERDD graph models, using GK to assess the graphlet topology (see Fig. 4). As with the BioGRID PPI networks, the STICKY model is all-round the best fit for the IID networks, with SF as a good second, even for the new mammalian species, while GEO and ER models are the worst. It is left as future work to determine whether these networks are also fit well by the HGG model.

Table 7. The 9 IID networks used in this study, in decreasing order of their average degree.

Name	Num nodes	Num edges	Density	Average degree
Yeast	6317	194700	0.00976	30.8
Guinea pig	14189	288179	0.00286	20.3
Cow	14783	297734	0.00272	20.1
Cat	14427	290367	0.00279	20.1
Dog	14512	287265	0.00273	19.8
Chicken	11833	227721	0.00325	19.2
Duck	11498	215383	0.00326	18.7
Turkey	10886	196798	0.00332	18.1
Fly	10310	110062	0.00207	10.7

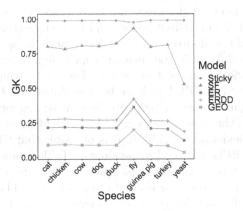

Fig. 4. We created 50 synthetic networks for each of the 5 models used to test against the IID PPI networks. The 9 PPI networks from IID have similar fits to the BioGRID ones. The STICKY model followed by SF are overall best fits for all the species while GEO and ER are the worst fits.

5   Discussion and Conclusion

We have performed a comprehensive evaluation of the fit of the largest BioGRID networks to several network models across a wide range of graphlet measures, and extended the analysis to far larger, machine-learned IID PPI networks. While there is significant disagreement between measures as to rank the models (as explicitly shown by the range of Pearson correlations between measures tabulated in Table 5), the gross conclusion is that the STICKY model is the overall best-fitting model to BioGRID PPI networks and that the HGG model is a good fit for all the PPI networks except DM and EC, and is a better model than STICKY on some PPI networks. STICKY's overall good performance is not surprising since it was specifically created to model the affinity of protein pairs to interact. The stickiness index is closely related not only to the degree-distribution, but to the *inter-node* topology of the network. It preserves graphlet structure better than pure theoretical models. STICKY has remained the best-fitting model throughout 10 years of increasingly voluminous PPI network data. However, while the STICKY model provides the best *description* of the networks, it provides little theoretical insight as to *why* they have the structure that they do. For that, we turn to other models to gain additional insight into the inter-connectivity and structure.

Previous studies [32] suggested that since yeast was geometric in structure (at the time the study was conducted and using the measures available at the time), the resulting network's degree distribution would be Poisson and hence, would have a peak at the mean degree. However, the current yeast network's degree distribution is far from Poisson, as the peak is at 1 (757 nodes with degree 1) but only 85 nodes have the mean degree of 25. We find that it is not GEO, but the HGG model and SF model that score well according to the best network measure as shown by the AUPR and AUROC curves in Fig. 2, suggesting that current PPI networks are partially scale-free and partially hyperbolic geometric in nature. As seen in Fig. 3 and Table 5, the measures have significant disagreement on the ordering of models from best to worst. The best-fitting models for PPI networks have changed as PPI data have been updated. We have also observed that the best-fitting model may depend on the comparison measure used for the assessment. Further, the 'best' fit may not be the *right* fit as we have seen with the graphlet count deviations in Table 6. It is plausible that STICKY or HGG *is* the right model for PPIs and the low p-values of Table 6 can be blamed not on the model, but on the noise and incompleteness of the data; this hypothesis will need to wait several more years for the data to catch up and be more thoroughly tested. Finally, the distinct disagreements between different graphlet measures depicted in Figs. 2 and 3 and Tables 5, and 6 suggest at present, the Graphlet Kernel may be the most reliable graphlet-based measure for network comparison since it has the greatest precision-recall.

References

1. Aldecoa, R., Orsini, C., Krioukov, D.: Hyperbolic graph generator. Comput. Phys. Commun. **196**, 492–496 (2015)
2. Barabási, A.L., Albert, R.: Emergence of scaling in random networks. Science **286**(5439), 509–512 (1999)

3. Barabási, A., Dezso, Z., Ravasz, E., Yook, Z.H., Oltvai, Z.N.: Scale-free and hierarchical structures in complex networks. In: Modeling of Complex Systems: Seventh Granada Lectures. AIP Conference Proceedings, vol. 661, pp. 1–16 (2003)
4. Bianconi, G., Pin, P., Marsili, M.: Assessing the relevance of node features for network structure. Proc. Nat. Acad. Sci. **106**(28), 11433–11438 (2009)
5. Chatr-Aryamontri, A., et al.: The BioGRID interaction database: 2017 update. Nucleic Acids Res. **45**(D1), D369–D379 (2017)
6. Davis, D., Yaveroğlu, Ö.N., Malod-Dognin, N., Stojmirovic, A., Pržulj, N.: Topology-function conservation in protein-protein interaction networks. Bioinformatics **31**(10), 1632–1639 (2015). https://doi.org/10.1093/bioinformatics/btv026
7. Davis, J., Goadrich, M.: The relationship between precision-recall and ROC curves. In: Proceedings of the 23rd International Conference on Machine Learning, pp. 233–240. ACM (2006)
8. Erdös, P., Rényi, A.: On random graphs. Publicationes Mathematicae **6**, 290–297 (1959)
9. Hayes, W., Sun, K., Pržulj, N.: Graphlet-based measures are suitable for biological network comparison. Bioinformatics **29**(4), 483–491 (2013)
10. Higham, D., Rašajski, M., Pržulj, N.: Fitting a geometric graph to a protein-protein interaction network. Bioinformatics **24**(8), 1093–1099 (2008)
11. Hočevar, T., Demšar, J.: A combinatorial approach to graphlet counting. Bioinformatics **30**(4), 559–565 (2014). https://doi.org/10.1093/bioinformatics/btt717
12. Janjić, V., Pržulj, N.: The topology of the growing human interactome data. J. Integr. Bioinform. **11**(2), 27–42 (2014)
13. Janjić, V., Sharan, R., Pržulj, N.: Modelling the yeast interactome. Sci. Rep. **4**, 4273 (2014)
14. Karlebach, G., Shamir, R.: Modelling and analysis of gene regulatory networks. Nat. Rev. Mol. Cell Biol. **9**(10), 770 (2008)
15. Kashtan, N., Itzkovitz, S., Milo, R., Alon, U.: Efficient sampling algorithm for estimating subgraph concentrations and detecting network motifs. Bioinformatics **20**(11), 1746–1758 (2004)
16. Kotlyar, M., Pastrello, C., Malik, Z., Jurisica, I.: IID 2018 update: context-specific physical protein-protein interactions in human, model organisms and domesticated species. Nucleic Acids Res. **47**(D1), D581–D589 (2018)
17. Krioukov, D., Papadopoulos, F., Kitsak, M., Vahdat, A., Boguná, M.: Hyperbolic geometry of complex networks. Phys. Rev. E **82**(3), 036106 (2010)
18. Kuchaiev, O., Pržulj, N.: Integrative network alignment reveals large regions of global network similarity in yeast and human. Bioinformatics **27**, 1390–1396 (2011). https://doi.org/10.1093/bioinformatics/btr127
19. Kuchaiev, O., Milenković, T., Memišević, V., Hayes, W., Pržulj, N.: Topological network alignment uncovers biological function and phylogeny. J. R. Soc. Interface **7**(50), 1341–1354 (2010). https://doi.org/10.1098/rsif.2010.0063
20. Luck, K., Sheynkman, G.M., Zhang, I., Vidal, M.: Proteome-scale human interactomics. Trends Biochem. Sci. **42**, 342–354 (2017)
21. Luo, F., Yang, Y., Chen, C.F., Chang, R., Zhou, J., Scheuermann, R.H.: Modular organization of protein interaction networks. Bioinformatics **23**(2), 207–214 (2006)
22. Malod-Dognin, N., Pržulj, N.: L-GRAAL: Lagrangian graphlet-based network aligner. Bioinformatics **31**(13), 2182–2189 (2015)
23. Mamano, N., Hayes, W.B.: SANA: simulated annealing far outperforms many other search algorithms for biological network alignment. Bioinformatics **33**, 2156–2164 (2017)

24. Milano, M., et al.: An extensive assessment of network alignment algorithms for comparison of brain connectomes. BMC Bioinform. **18**(6), 235 (2017)

25. Milenković, T., Lai, J., Pržulj, N.: GraphCrunch: a tool for large network analyses. BMC Bioinform. **9**, 70 (2008)

26. Milenković, T., Pržulj, N.: Uncovering biological network function via graphlet degree signatures. Cancer Inform. **6**, 257–273 (2008)

27. Milenković, T., Ng, W.L., Hayes, W., Pržulj, N.: Optimal network alignment with graphlet degree vectors. Cancer Inform. **9**, 121–137 (2010). https://doi.org/ 10.4137/CIN.S4744. http://www.la-press.com/optimal-network-alignment-with-graphlet-degree-vectors-article-a2141

28. Penrose, M.: Random Geometric Graphs. Oxford Studies in Probability. Oxford University Press, Oxford (2003)

29. Petit, J., Kavelaars, J., Gladman, B., Loredo, T.: Size distribution of multikilometer transneptunian objects. In: The Solar System Beyond Neptune, pp. 71–87 (2008)

30. Pinkert, S., Schultz, J., Reichardt, J.: Protein interaction networks-more than mere modules. PLoS Comput. Biol. **6**(1), e1000659 (2010)

31. Pržulj, N.: Biological network comparison using graphlet degree distribution. Bioinformatics **20**, e177–e183 (2007)

32. Pržulj, N., Corneil, D.G., Jurisica, I.: Modeling interactome: scale-free or geometric? Bioinformatics **20**(18), 3508–3515 (2004). https://doi.org/10.1093/ bioinformatics/bth436. http://bioinformatics.oxfordjournals.org/content/20/18/ 3508.abstract

33. Pržulj, N., Higham, D.: Modelling protein-protein interaction networks via a stickiness index. J. R. Soc. Interface **3**(10), 711–716 (2006)

34. Pržulj, N., Kuchaiev, O., Stevanović, A., Hayes, W.: Geometric evolutionary dynamics of protein interaction networks. In: Pacific Symposium on Biocomputing (2010)

35. Pržulj, N., Kuchaiev, O., Stevanović, A., Hayes, W.: Geometric evolutionary dynamics of protein interaction networks. In: Proceedings of the 2010 Pacific Symposium on Biocomputing (PSB), 4–8 January 2010, Big Island, Hawaii (2010)

36. Pržulj, N., Milenković, T.: Computational methods for analyzing and modeling biological networks. In: Chen, J., Lonardi, S. (eds.) Biological Data Mining. CRC Press (2009, To appear)

37. Pržulj, N.: Biological network comparison using graphlet degree distribution. Bioinformatics **23**(2), e177–e183 (2007)

38. Rito, T., Wang, Z., Deane, C.M., Reinert, G.: How threshold behaviour affects the use of subgraphs for network comparison. Bioinformatics **26**(18), i611–i617 (2010). https://doi.org/10.1093/bioinformatics/btq386

39. Salathé, M., May, R.M., Bonhoeffer, S.: The evolution of network topology by selective removal. R. Soc. Interface **2**, 533–536 (2005)

40. Shen-Orr, S., Milo, R., Mangan, S., Alon, U.: Network motifs in the transcriptional regulation network of Escherichia coli. Nat. Genet. **31**(1), 64–68 (2002)

41. Shervashidze, N., Vishwanathan, S., Petri, T., Mehlhorn, K., Borgwardt, K.: Efficient graphlet kernels for large graph comparison. In: Artificial Intelligence and Statistics, pp. 488–495 (2009)

42. Sole, R.V., Manrubia, S.C., Benton, M., Bak, P.: Self-similarity of extinction statistics in the fossil record. Nature **388**(6644), 764 (1997)

43. Vázquez, A., Flammini, A., Maritan, A., Vespignani, A.: Modeling of protein interaction networks. Complexus **1**(1), 38–44 (2003)

44. Vijayan, V., Saraph, V., Milenković, T.: MAGNA++: maximizing accuracy in global network alignment via both node and edge conservation. Bioinformatics **31**, 2409–2411 (2015)
45. Vishveshwara, S., Brinda, K., Kannan, N.: Protein structure: insights from graph theory. J. Theor. Comput. Chem. **1**(01), 187–211 (2002)
46. Wang, Z., Zhang, J.: In search of the biological significance of modular structures in protein networks. PLoS Comput. Biol. **3**(6), e107 (2007)
47. Yaveroğlu, N., et al.: Revealing the hidden language of complex networks. Sci. Rep. **4**, 4547 (2014)
48. Yaveroğlu, Ö.N., Milenković, T., Pržulj, N.: Proper evaluation of alignment-free network comparison methods. Bioinformatics **31**(16), 2697–2704 (2015)

Graph-Theoretic Partitioning of RNAs and Classification of Pseudoknots

Louis Petingi[1,2](✉) and Tamar Schlick[3,4]

[1] Computer Science Department,
College of Staten Island (CUNY), New York City, USA
louis.petingi@csi.cuny.edu
[2] City University of New York Graduate Center, New York City, USA
[3] Department of Chemistry,
Courant Institute of Mathematical Sciences, NYU, New York City, USA
schlick@nyu.edu
[4] NYU-East China Normal University Center for Computational Chemistry,
New York University Shanghai, Shanghai, China

Abstract. Dual graphs have been applied to model RNA secondary structures with pseudoknots, or intertwined base pairs. In a previous work, a linear-time algorithm was introduced to partition dual graphs into maximal topological components called blocks and determine whether each block contains a pseudoknot or not. This characterization allowed us to efficiently isolate smaller RNA fragments and classify them as pseudoknotted or pseudoknot-free regions, while keeping these substructures intact. In this paper we extend the partitioning algorithm by classifying a pseudoknot as either recursive or non-recursive. A pseudoknot is recursive if it contains independent regions or fragments. Each of these regions can be also identified by the modified algorithm, continuing with our current research in the development of a library of building blocks for RNA design by fragment assembly. Partitioning and classification of RNAs using dual graphs provide a systematic way for study of RNA structure and prediction.

Keywords: Graph theory · RNA secondary structures · Partitioning · Bi-connectivity · Pseudoknots

1 Introduction

Let $G = (V, E)$ be undirected graph composed of by a finite set of *vertices* V and a set E of unordered pair of vertices called *edges*, where each edge represents a relation between two vertices.

Our RNA analysis is based on dual graphs, introduced in 2003 by Gan et al. [7], to model RNA secondary structures (2D). The 2D elements of RNA molecules consist of double-stranded (stem) regions defined by base pairing such as Adenine-Uracil, Guanine-Cytosine, Guanine-Uracil, and single stranded loops; stems and loops are mapped to the vertices and edges of the corresponding dual

© Springer Nature Switzerland AG 2019
I. Holmes et al. (Eds.): AlCoB 2019, LNBI 11488, pp. 68–79, 2019.
https://doi.org/10.1007/978-3-030-18174-1_5

graph, respectively (later we present an alternative definition of dual graphs). Dual graphs can represent complex RNA structures called pseudoknots (*PKs*); these structures involve an intertwining of two-base-paired regions of the RNA and are common elements in many biologically relevant RNAs.

In [19,20] a linear-time partitioning algorithm was introduced based on the dual graph representation of RNA 2D. This algorithm partitions a dual graph into connected components called *blocks* and then determines whether each block contains a pseudoknot or is a regular region. Thus our procedure provides a systematic approach to partition an RNA 2D, into smaller classified regions, while providing a topological perspective for the analysis of RNAs.

Pseudoknots can be classified into two main groups: *recursive* and *non-recursive* pseudoknot [9,23]. The former is distinguished from the latter because it contains an internal pseudoknotted or regular region that does not intertwine with external stems within the PK; in this work, the original algorithm is extended to classify PKs into these two main categories. In addition, as a recursive PK comprises independent regions or fragments, our modified algorithm can also identify each of these regions, to be later cataloged and applied in the analysis of RNAs with pseudoknots.

In the next section, we present background material and definitions relevant to this paper, and we review the partitioning algorithm introduced in [19,20], as well as its applications, as for example the development of a library of building blocks for RNA design by fragment assembly [13]. Following this line of research, in Sect. 3 it is shown how the partitioning approach can be extended so if a block contains a pseudoknot, then it can be classified as either recursive or non-recursive; in the case the PK is recursive, the algorithm can also identify each independent region. We summarize the findings and outline new directions in Sect. 4. An Appendix section includes computational tests performed by the modified algorithm, on some RNA's motifs.

2 Background

2.1 Biological and Topological Definitions

In 2003, Gan et al. [7] introduced *tree* and *dual* graph-theoretic representations of RNA 2D motifs in a framework called RAG (RNA-As-Graphs) [5,8,12,16]. A pseudoknot is an intertwining of two-based-paired regions (stems) of an RNA (see for example Fig. 1).

The partition algorithm is based on topological properties of graphs, suggests an alternative way to look at the problem of detection and classification of PKs and of general RNAs. As base pairing in PKs is not well-nested, making the presence of PKs in RNA sequences more difficult to predict by the more classical dynamic programming [3] and context-free grammars standard methods [2].

Following (Kravchenko [17]), we define our biological variables as follows.

Definition 1. *General terms:*

a. *RNA primary structure: a sequence of linearly ordered bases* $x_1, x_2, \ldots, x_r$, *where* $x_i \in \{A, U, C, G\}$.

b. *canonical base pair: a base pair* $(x_i, x_j) \in \{(A, U), (U, A), (C, G), (G, C), (G, U), (U, G)\}$.

c. *RNA secondary structure without pseudoknot - or regular structure, encapsulated in the region* $(i_0, \ldots, k_0)$: *an RNA 2D structure in which no two base pairs* $(x_i, x_j), (x_l, x_m)$, *satisfy* $i_0 \leq i < l < j < m \leq m_0$ *(i.e., no two base pairs intertwined).*

d. *a base pair stem: a tuple* $(x_i, x_{i+1}, \ldots, x_{i+r}, x_{i+(r+1)}, \ldots, x_{j-1}, x_j)$ *in which* (x_i, x_j), $(x_{i+1}, x_{j-1}), \ldots, (x_{i+r}, x_{i+(r+1)})$ *form base pairs.*

e. *loop region: a tuple* $(x_1, x_2, \ldots, x_r)$ *in which* $\forall_{i \leq j \leq r}(x_i, x_j)$ *does not form a base pair.*

f. *a pseudoknot encapsulated in the region* $(i_0, \ldots, k_0)$: *if* $\exists l, m, (i_0 < l < m < k_0)$ *such that* (x_{i_0}, x_m) *and* (x_l, x_{k_0}) *are base pairs (i.e., at least two base pairs intertwined).*

A graphical representation is a natural way to describe an RNA 2D structure (see Fig. 1(a), (b)), in which the x-axis is labeled according to the primary linearly ordered sequence of bases (Definition 1a), and a stem (Definition 1d) is represented by arcs connecting base pairs. A region on the x-axis between the end-points of the arcs representing stems is called a *segment*.

A dual graph can be defined from the graphical representation of an RNA 2D structure as follows (Fig. 1).

Definition 2. *The dual graph is defined by mapping stems and the segments between stems (x-axis), of the graphical representation of an RNA 2D structure, to the vertices and edges of the dual graph, respectively.*

In the next section we present our partitioning approach as of a dual graph G, into subgraphs $G' \subseteq G$, called blocks.

2.2   Graph Partitioning Algorithm

The graph-theoretic partitioning algorithm is based on identifying *articulation points* of the dual graph representation of an RNA 2D. An articulation point is a vertex of a graph whose deletion disconnects a graph or an isolated vertex remains.

We need to define the following.

Definition 3. *Connectivity*

a. *A vertex v is an articulation point or cut-vertex if $G - v$ results in a disconnected graph (i.e., at least two connected components remain) or an isolated vertex remains.*

b. *A connected component is non-separable if it does not have an articulation point (or cut-vertex). Please note that single edges or isolated points are non-separable.*

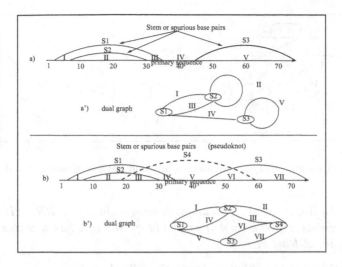

Fig. 1. Graphical and dual graph representations of an RNA 2D structure. (a) graphical representation of a pseudoknot-free RNA primary sequence and embedded stems or base pairs; (a′) corresponding dual graph representation. (b) graphical representation of a pseudoknotted RNA 2D structure; (b′) corresponding dual graph. This figure was originally depicted in [20].

c. A block is a maximal (edge-wise) non-separable graph.
d. An edge-set X is an edge-disconnecting set if the removal of X from G results in a disconnected graph. The edge-connectivity of a graph $\lambda(G)$ is the size of a minimum edge-disconnecting of G.
e. The degree of a vertex v of G is the number of $d_G(v)$ is the number of edges incident at v.

Articulation points allow us to identify blocks (see Fig. 2); since a block is a maximally non-separable component, a pseudoknot cannot be then contained in two different blocks. Thus identification of these block components allows us to isolate pseudoknots (as well as pseudoknot-free blocks), without breaking their structural properties.

An algorithm for identifying (bi-connected) block components in a graph was introduced by Hopcroft and Tarjan [11], and runs in linear computational time.

A *hairpin* loop occurs when two regions of the same strand, usually complementary in nucleotide sequence when read in opposite directions, base-pair to form a double helix that ends in an unpaired loop. A self-loop in the dual graph, i.e., an edge having the same vertex as the end-points, represents a hairpin, and as it does not connect two different vertices (i.e., stems), it is formally deleted from the dual graph.

From Definition 1c, an RNA 2D structure is a regular-region (pseudoknot-free) and encapsulated in a region $(i_0, \ldots, k_0)$, if no two base pairs $(x_i, x_j), (x_l, x_m)$, satisfy $i < l < j < m$, $i_0 \leq i, j, l, m \leq m_0$, otherwise the region is a pseudoknot; this definition yields the following main result.

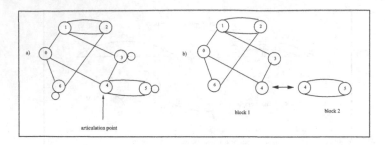

Fig. 2. Identification of (a) articulation points and (b) partitioning of a dual graph.

Corollary 4. *[19, 20] Given a dual graph representation of RNA 2D structure, a block represents a pseudoknot if and only if the block has a vertex of degree (Definition 3e) at least 3.*

The partitioning algorithm performs the following steps.

1. *Partition the dual graph into blocks by application of Hopcroft and Tarjan's algorithm.*
2. *Analyze each block to determine whether contains a vertex of degree at least 3. If that is the case then the block contains a pseudoknot, according to Corollary 4. If not then the block represents a pseudoknot-free structure.*

Consider as an example the dual graph shown in Fig. 2. This graph is decomposed into 2 blocks. According to Corollary 4, block 1 is a pseudoknot as it has a vertex of degree at least 3, while block 2, a cycle, corresponds to a regular region.

Our partitioning algorithm was applied recently [13] to analyze the modular units of RNAs for a representative database of experimentally determined RNA structures and to develop a library of building blocks for RNA design by fragment assembly, as done recently for tree graphs, along with supporting chemical mapping experiments [14]. Among the 22 frequently occurring motifs we found for known RNAs up to 9 vertices, 15 contain pseudoknots [13]. Thus, further classification of the pseudoknotted RNAs could help in cataloging and applications to RNA design. Another application of the partitioning algorithm to small and large units of ribosomal RNAs of various prokaryotic and eukaryotic organisms helped identify common subgraphs and ancestry relationships [13].

In the next section we extend our algorithm to classify PKs as either recursive or non-recursive; the algorithm can also identify each recursive region.

3   Classification of Pseudoknots as Either Recursive or Non-recursive and Identification of Each Recursive Region

The RNA 2D dual graph and graphical representations depicted in this section are based upon New York University's *RAG*-database [12], and *R-Chie* visualization software [18], respectively.

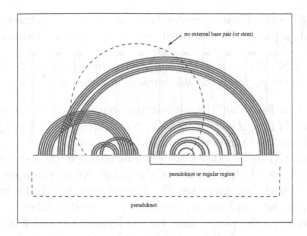

Fig. 3. Recursive pseudoknot.

The definition of a recursive pseudoknot follows the one of Wong et al. [23]. Let $A = x_1x_2...x_m$ be a sequence of linearly ordered bases, and M be the 2D of A. M is represented as a set of base pair positions, i.e., $M = \{(i,j)|1 \leq i < j \leq m, (x_i, x_j)$ is a base pair$\}$. Let $M_{x,y} \subseteq M$ be the set of base pairs within the subsequence $x_1x_2...x_m$, $1 \leq x < y \leq m$, i.e., $M_{x,y} = \{(i,j) \in M|x \leq i < j \leq y\}$, with $M = M_{1,m}$.

Definition 5. $M_{x,y}$ is a recursive pseudoknot if $M_{x,y}$ is a pseudoknot (see Definition 1f), and $\exists a_1, b_1, \ldots, a_s, b_s, (x < a_1 < b_1 < \ldots < a_s < b_s < y)$ that satisfy the followings.
Each M_{a_i,b_i} is called a recursive region.

- M_{a_i,b_i}, for $1 \leq i \leq s$, is a recursive pseudoknot.
- For each $M_{a_i,b_i}, 1 \leq i \leq s$, there does not exist a base pair $(i,j) \in M$ that $i \in [a_i, b_i]$ but $j \notin [a_i, b_i]$, or $i \notin [a_i, b_i]$ but $j \in [a_i, b_i]$.
- $M_{x,y} - \cup_{i=1}^s M_{a_i,b_i}$ is either a regular structure or a pseudoknot.

A recursive pseudoknot is a pseudoknot $M_{x,y}$ that contains a pseudoknotted or regular region $M_{a,b}$, and there does not exist a base pair (c, d), such that x_d is a base of $M_{a,b}$ and x_c is a base of M external to $M_{a,b}$ (see Fig. 3). Here we are assuming that $M_{a,b}$ is contained in $M_{x,y}$, that is, $x < a < b < y$.

Wong et al. definitions [23] also incorporated the concepts of *standard* and *non-standard* pseudoknots; however it is not within the scope of this work to study them from the dual graph representation perspective.

A graph is *Eulerian* if there exist a trail with no repetition of edges from a vertex v_0 of G, ending at vertex v_k, covering all the edges of the topology; if $v_0 = v_k$ then the graph is an *Eulerian cycle* (see [10], p. 64). Dual graph representations of general RNA 2D structures, and specifically of PKs, can be easily shown to be Eulerian graphs from Definition 2. By starting from the origin on the x-axis of the graphical representation and traversing to the right, a unique trail in its dual graph can be described, where all edges are covered.

Lemma 6. *[19, 20] The dual graph representations of RNA 2D structures and of PKs are Eulerian by following the primary sequence of bases.*

As depicted in Fig. 1(b), the alternating sequence of stems and segments $\{S_1, I, S_2, II, S_4, III, S_2, IV, S_1, V, S_3, VI, S_4, VII, S_3\}$ of the graphical representation (b) forms an Eulerian trail in its dual graph (b').

A pseudoknotted block can be classified as recursive by just calculating the edge-connectivity (see Definition 3d) of the block. As an example consider the *Hepatitis Delta Virus Ribozyme* (see Fig. 4), necessary for viral replication. The stem labeled 4 in the graphical representation (or vertex labeled 4 in the dual graph) is attached to the pseudoknot by the segments a and b in its graphical representation, or edges labeled a and b in the dual graph representation. It is clear that if the PK is recursive then the edge-connectivity of the pseudoknotted block must be 2. However it is not obvious that the converse is necessarily true, that is, if the pseudoknotted block has edge-connectivity 2 then it is recursive. The following Lemma settles this question.

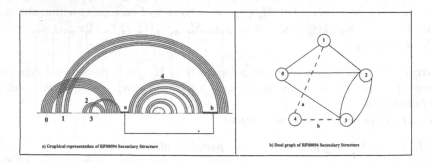

Fig. 4. Hepatitis Delta Virus Ribozyme secondary structure. (a) Graphical representation. (b) Dual graph representation.

Lemma 7. *The dual graph representation of a pseudoknotted block is recursive if and only if the block has edge-connectivity 2.*

Proof. If the block $M_{x,y}$ is a recursive pseudoknot then it contains an internal region $M_{a,b}$ with $x < a < b < y$ according the aforementioned definition. As there does not exist a base pair (c, d) in which x_d is a base in the internal region and x_c is a base of the pseudoknot outside this internal region, then $M_{a,b}$ must be adjacent to the remaining of the PK in the graphical representation by two segments, or equivalently, by two edges in the dual graph of the pseudoknot (see Fig. 4).

Conversely suppose that the dual graph of a PK, $G = (V, E)$, has edge-connectivity 2 and let $E' = \{e_1, e_2\}$ represent a minimum size disconnecting set. As G is connected then deletion of E' from G will result in exactly two connected components, $G_1 = (V_1, E_1)$ and $G_2 = (V_2, E_2)$ (see Fig. 5a). From Lemma 6 one knows that there exist an Eulerian path in G following the primary sequence in linear order, starting from the origin on the x-axis of the graphical representation and traversing to the right. Let the Eulerian path $P = P_1.e_1.P_2.e_2.P_3$ where P_1 starts at the initial base x_1 (w.l.o.g. we assume that x_1 is in G_1) (see Fig. 5b). It is the case that P_2 must cover all the edges of G_2 (following an ordered sequence of bases) because when P reaches e_2 to continue with P_3, P can not go back to G_2 again as e_1 and e_2 were already used by the Eulerian path. Therefore G_2 is a region composed of all edges (and vertices) corresponding to an ordered sequence of bases, that is, G_2 is a well-defined region within the pseudoknotted block G. □

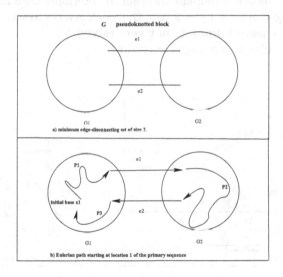

Fig. 5. Minimum disconnecting set of size 2. (a) Two connected subgraphs. (b) Eulerian path covering all the edges of the dual graph and following the primary sequence.

The edge-connectivity of a graph $G = (V, E)$ can be determined in polynomial time in order $(|V||E|^2)$ using the max-flow min-cut theorem of network flows by Edmond and Karp [4], or Ford and Fulkerson [6].

As it is shown in the proof of by Lemma 7, we can also delete each pair of edges and determine if the graph is disconnected using Depth-First-Search [10] in time $(|E|^3)$; this variation allows us to find every internal recursive region of the recursive pseudoknot if such pair of edges exist. For example if edges a and b of Hepatitis Delta Virus Ribozyme 2D dual graph representation (Fig. 4) are deleted, then vertex labeled 4 corresponding to the stem labeled 4 of the graphical representation will be isolated. Thus the disconnecting set $\{a, b\}$ uniquely identifies a recursive region (i.e., stem labeled 4).

Corollary 4 and Lemma 7 yield the following partitioning and classification of pseudoknots algorithm.

Algorithm for Partitioning and Classification of PKs

i. Input dual graph $G = (V, E)$ as the Adjacency Matrix, of a RNA 2D.
ii. Output partitioning of the RNA 2D into recursive PK, non-recursive PK, and regular regions.

  1. *Partition the dual graph into blocks by application of Hopcroft and Tarjan's algorithm;*
  2. *Analyze each block to determine whether each contains a vertex of degree at least 3;*
  3. **IF** the block has a vertex of degree ≥ 3 *then* the block is a **pseudoknot;**
    • Apply max-flow min-cut theorem to determine edge-connectivity;
    • *if* edge-connectivity $= 2$ then the block is a **recursive pseudoknot;**
    else the **pseudoknot is not recursive;**
  4. **ELSE** the **block is a regular region;**.

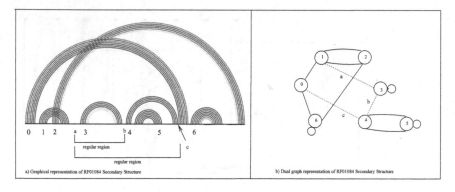

Fig. 6. A *tRNA-like-structure*. (a) Graphical representation. (b) Dual graph representation.

As another example consider a *tRNA-like-structure* [1], linked to regulation of plant virus replication (see Fig. 6). Note that the vertex labeled 4 is an articulation point, therefore this dual graph will be partitioned into two blocks, one is a pseudoknot, because it contains a vertex of degree 3 or greater (self-loops are deleted in the analysis), while the other block is a regular region. The pseudoknotted bock can be classified as recursive because it has edge-connectivity 2. In addition every disconnecting set of size 2 represents a recursive internal region of the PK.

As an example of a non-recursive pseudoknot consider the *Translational repression of the Escherichia coli alpha operon mRNA* [22], illustrated in Fig. 7. The dual graph representation of this motif 2D has edge-connectivity 3, thus it is not a recursive PK.

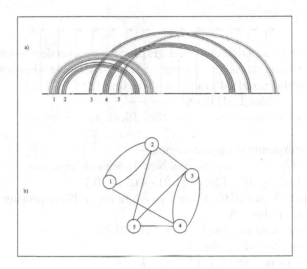

Fig. 7. *Translational repression of the Escherichia coli alpha operon mRNA.* (a) Graphical representation; (b) Dual graph representation.

The Appendix illustrates the output generated by the modified algorithm when is run on some of the aforementioned motifs. The algorithm is written in C++ and is archived for public use [21].

4   Conclusions and Ongoing Work

We have extended our partitioning algorithm of the dual graph representation of RNA 2D structures into maximal non-separable components called blocks, to classify pseudoknots as either recursive or non-recursive. In [19,20] it was shown that an RNA 2D contains a pseudoknot if and only if the dual graph representation has a block in which one of the vertices is of degree 3 or larger. This paper showed that a pseudoknotted block is recursive if and only if the block has edge-connectivity 2. Moreover each disconnecting set of size 2 represents an internal recursive region of the pseudoknot allowing further classification of modular units for RNA design. These results also offer an alternative and simple way to visualize and classify PKs based on graph theoretical properties, allowing a systematic analysis of RNAs.

With our recent extension of our graph growing algorithm to generate dual graph libraries of possible RNA motifs, thousands more potential graphs were analyzed and classified [15]. The modified algorithm will be useful for these motifs for further studies of RNA structure and design.

Acknowledgments. We would like to thank first the referees for improving the content of the paper. NIGMS support from award R35GM122562 to T. S. is gratefully acknowledged. The work of L. P. was supported by PSC-CUNY Award # 61249-00-49 of the City University of New York.

Appendix

Let (a, b) represents an edge of a dual graph with end-vertices a and b.

We are next illustrating the output of the partitioning algorithm tested on the tRNA-like-structure dual graph (see Fig. 6).

———————— Motif :RF01084 ————————————

==================== New Block =======

(4,5) - (4,5) -

—- this block represents a regular-region —-

==================== New Block =======

(4,0) - (3,4) - (1,3) - (6,0) - (2,6) - (1,2) - (1,2) - (0,1) -

removed edges (2,6) and (6,0), these two edges are a disconnecting set:

The block is a recursive PK.

————— Summary information for Motif :RF01084 ————————

————— Total number of blocks: 2

————— number of non-recursive PK blocks: 0

————— number of recursive PK blocks: 1

————— number of regular blocks : 1

References

1. Barends, S., Rudinger-Thirion, J., Florentz, C., Giegé, R., Pleij, C., Kraal, B.: tRNA-Like structure regulates translation of brome mosaic virus RNA. J. Virol. **78**(8), 4003–4010 (2004)
2. Condon, A., Davy, B., Rastegari, B., Zhao, S., Tarrant, F.: Classifying RNA pseudoknotted structures. Theor. Comput. Sci. **320**(1), 35–50 (2004)
3. Dirks, R.-M., Pierce, N.-A.: A partition function algorithm for nucleic acid secondary structure including pseudoknots. J. Comput. Chem. **24**(13), 1664–1677 (2003)
4. Edmonds, J., Karp, R.-M.: Theoretical improvements in algorithmic efficiency for network flow problems. J. ACM **19**(2), 248–264 (1972)
5. Fera, D., et al.: RAG: RNA-As-Graphs web resource. BMC Bioinform. **5**, 88 (2004)
6. Ford, L.R., Fulkerson, D.R.: Maximal flow through a network. Can. J. Math. **8**(1956), 399–404 (1956)
7. Gan, H.-H., Pasquali, S., Schlick, T.: Exploring the repertoire of RNA secondary motifs using graph theory; implications for RNA design. Nucleic Acids Res. **31**(11), 2926–2943 (2003)
8. Gan, H.-H., et al.: RAG: RNA-As-Graphs database-concepts, analysis, and features. Bioinformatics **20**(8), 1285–1291 (2004)
9. Han, B., Dost, B., Bafna, V., Zhang, S.: Structural alignment of pseudoknotted RNA. J. Comput. Biol. **15**(5), 489–504 (2008)

10. Harary, F.: Graph Theory. Addison-Wesley, Boston (1969)
11. Hopcroft, J., Tarjan, R.: Efficient algorithms for graph manipulation. Communi. ACM **16**(6), 372–378 (1973)
12. Izzo, J.-A., Kim, N., Elmetwaly, S., Schlick, T.: RAG: an update to the RNA-As-Graphs resource. BMC Bioinform. **12**, 219 (2011)
13. Jain, S., Bayrak, C.-S., Petingi, L., Schlick, T.: Dual graph partitioning highlights a small group of pseudoknot-containing RNA submotifs. Genes **9**(8), 371 (2018)
14. Jain, S., Ramos, S.-B., Laederach, A., Schlick, T.: A pipeline for computational design of novel RNA-like topologies. NAR **46**(14), 7040–7051 (2018)
15. Jain, S., Saju, S., Petingi, L., Schlick, T.: An extended dual graph library and partitioning algorithm applicable to pseudoknotted RNA structures. Methods (2019). https://doi.org/10.1016/j.ymeth.2019.03.022. Accessed 1 Apr 2019
16. Kim, N., Petingi, L., Schlick, T.: Network theory tools for RNA modeling. WSEAS Trans. Math. **12**(9), 941–955 (2013)
17. Kravchenko, A.: Predicting RNA secondary structures including pseudoknots. University of Oxford internal report (2009)
18. Lai, D., Proctor, J.-R., Zhu, J.-Y., Meyer, I.-M.: R-CHIE: a web server and R package for visualizing RNA secondary structures. Nucleic Acids Res. **40**(12), e95 (2012). e-RNA: https://www.e-rna.org/r-chie/rfam.cgi. Accessed 4 Jan 2019
19. Petingi, L., Schlick, T.: Partitioning RNAs into pseudonotted and pseudoknot-free regions modeled as dual graphs. q-bio.QM, arXiv:1601.04259 (2016). http://arxiv.org/abs/1601.0425. Accessed 4 Oct 2018
20. Petingi, L., Schlick, T.: Partitioning and classification of RNA secondary structures into pseudonotted and pseudoknot-free regions using a graph-theoretical approach. IAENG Int. J. Comput. Sci. **44**(2), 241–246 (2017)
21. Petingi, L.: Dual graph partitioning code. https://github.com/Louis-Petingi/Partition-Algorithm-3. Accesed 3 Jan 2019
22. Schlax, P.-J., Xavier, K.-A., Gluick, T.-C., Draper, D.-E.: Translational repression of the Escherichia coli alpha operon mRNA: importance of an mRNA conformational switch and a ternary entrapment complex. J. Biol. Chem. **276**(42), 38494–38501 (2001)
23. Wong, T.-K., Lam, T.-W., Sung, W.-K., Cheung, B.-W., Yiu, S.-M.: Structural alignment of RNA with complex pseudoknot structure. J. Comput. Biol. **18**(1), 97–108 (2011)

PathRacer: Racing Profile HMM Paths on Assembly Graph

Alexander Shlemov and Anton Korobeynikov(✉)

Center for Algorithmic Biotechnology, Saint Petersburg State University,
Saint Petersburg 198504, Russia
a.korobeynikov@spbu.ru

Abstract. Recently large databases containing profile Hidden Markov Models (pHMMs) emerged. These pHMMs may represent the sequences of antibiotic resistance genes, or allelic variations amongst highly conserved housekeeping genes used for strain typing, etc. The typical application of such a database includes the alignment of contigs to pHMM hoping that the sequence of gene of interest is located within the single contig. Such a condition is often violated for metagenomes preventing the effective use of such databases.

We present PATHRACER—a novel standalone tool that aligns profile HMM directly to the assembly graph (performing the codon translation on fly for amino acid pHMMs). The tool provides the set of most probable paths traversed by a HMM through the whole assembly graph, regardless whether the sequence of interested is encoded on the single contig or scattered across the set of edges, therefore significantly improving the recovery of sequences of interest even from fragmented metagenome assemblies.

Keywords: Profile HMM · Graph alignment ·
Set of most probable paths

Availability: http://cab.spbu.ru/software/pathracer/.

1 Introduction

The recent advances of metagenomics quickly revealed scalability issues arising from the amount of raw sequencing data necessary to describe complex microbial communities. Even more, due to many inherit problems and challenges the assembly of metagenomic data remains a non-trivial task, thus slowing down biological discoveries [13]. Still, in the majority of cases the interest of researchers lies around the recovery of certain important genes this way the use of additional information about these genes might be crucial for the full gene sequence recovery even from very fragmented assemblies.

The typical way to represent the sequence of the particular gene family is via so-called profile Hidden Markov Models [5]. Recently large databases such

© Springer Nature Switzerland AG 2019
I. Holmes et al. (Eds.): AlCoB 2019, LNBI 11488, pp. 80–94, 2019.
https://doi.org/10.1007/978-3-030-18174-1_6

as Pfam [8] or NCBIfam-AMR [1] containing thousands of pHMMs emerged. These pHMMs may represent the sequences of antibiotic resistance genes, or allelic variations amongst highly conserved housekeeping genes.

The recent releases of XANDER [19] and MEGAGTA [11] tools opened the possibility of the gene-targeted metagenomics assemblies, where the trained Hidden Markov Model is used to guide the traversal of de Bruijn graph. Such an approach gives obvious advantage over other assembly methods. Still, these tools might be somehow non-trivial to use, for example, XANDER forces a user to build both forward and reverse HMMs and MEGAGTA requires original sequences the pHMM was built from. All these requirements makes the usage of these tools non-straightforward as one cannot simply download the HMM from the database and use it straight away. Even more, both tools includes their own genome assembler engines essentially ignoring the recent progress made in the field of genome and metagenome assemblies.

PathRacer is a novel standalone tool that aligns profile HMM directly to the assembly graph produced by modern assemblers in standard GFA format[1]. This way PathRacer could utilize all the features and improvements (e.g. hybrid or mate-pair assemblies) provided by state-of-the-art assemblers. Below we describe algorithmic approaches used in PathRacer and showcase them over several datasets.

2   Methods

2.1   General Definitions

Let *Sequence graph* G be a directed graph where each vertex (called *position*) is labelled by a letter from the given alphabet Σ. Thus, each path in G corresponds to some string in Σ^*. Denote by $V(G)$ and $E(G)$ the sets of vertices and edges of G correspondingly.

pHMM graph is a profile HMM in HMMER 3 *Plan 7* format [5,6] on Σ. Profile HMM can be viewed as weighted directed graph with vertices called *states* and edges called *transitions*. A particular state could be of one of the following types: I *(insertion)*, M *(match or mismatch)*, D *(deletion)*; in addition to this there are also two special states denoted by START and END (see Fig. 1). I and M states together are called *emissive states* or *emitters*. Each emissive state E contains its own emission probabilities $\mathcal{E} = (E, \cdot)$ that define a probability distributions on Σ. Edges weights denoted as $\mathcal{T}(\cdot, \cdot)$ are transition probabilities; in each vertex sum of its outgoing transitions equals to 1. Note that pHMM graph is *almost* acyclic: the only allowed cycles are simple loops from each I state to itself *(I-loops)*. Therefore, pHMM states could be topologically sorted.

Background distribution $\mathcal{B}$ is a discrete probability distribution on Σ. This distribution is an *a priori* distribution of letters in the model. See [6] for more information about background distribution and its role.

---

[1] So far only GFA from de Bruijn graph assemblers like SPADES and MEGAHIT is supported, but we will address this restriction in the next PathRacer versions.

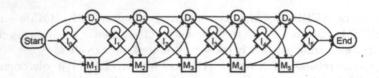

Fig. 1. Plan 7 profile hidden Markov model (pHMM) scheme

For string $(L_1, \ldots, L_n) \in \Sigma^*$ *HMM alignment* is a path in the pHMM graph that contains exactly n emitters. If the path begins from START and ends with END the alignment is called *global*, otherwise the alignment is called *local*. For the sake of simplicity we will consider only the problem of global alignment search. Note that the local alignments could be turned into the global ones by a simple transformation of pHMM graph [6]: insert additional transitions from START to all emitters and from all emitters to END.

Furthermore, the global alignment is fully defined by the series of its emitters $(E_1, \ldots, E_n)$. Indeed, the "missed" intermediate D states in the alignment path could be unambiguously reconstructed due to a particular structure of pHMM.

For the given pHMM graph, background distribution $\mathcal{B}$, string $\mathcal{L} = (L_1, \ldots L_n)$, and its global alignment $\mathcal{A} = (E_1, \ldots, E_n)$ alignment score $Score(\mathcal{L}, \mathcal{A})$ is defined as:

$$Score(\mathcal{L}, \mathcal{A}) = \log(\mathcal{T}(\text{START}, E_1)) + \log(\mathcal{E}(E_1, L_1) - \log(\mathcal{B}(L_1))$$
$$+ \log(\mathcal{T}(E_1, E_2)) + \cdots + \log(\mathcal{E}(E_n, L_n) - \log(\mathcal{B}(L_n)) + \log(\mathcal{T}(E_n, \text{END})),$$

where $\mathcal{T}(E_i, E_{i+1})$ may include intermediate D states $D_i^{(1)}, D_i^{(2)}, \ldots, D_i^{(T)}$:

$$\log(\mathcal{T}(E_i, E_{i+1})) = \log(\mathcal{T}(E_i, D_i^{(1)})) + \cdots + \log(\mathcal{T}(D_i^{(T)}, E_{i+1})).$$

This score is essentially

$$\log \frac{Prob(\text{"string produced by this sequence of HMM states"})}{Prob(\text{"string is randomly obtained from the background distribution"})}$$

and is known as *log-odds score* [6].

For given HMM graph and background define the alignment score $Score(\mathcal{L})$ for the string $\mathcal{L} = (L_1, \ldots, L_n)$ as the best score along all its possible global alignments $\mathcal{A} = (E_1, \ldots, E_n)$.

We consider the following problem of graph alignment to profile HMM: for given sequence graph G, profile HMM $pHMM$, background distribution $\mathcal{B}$, and integer $k > 0$, among all the paths in G obtain k different paths with the highest possible scores. We will call this problem *Top(k) path problem* throughout the article.

It is well known that for the one string the best score alignment could be found by Viterbi dynamic programming algorithm [18]. Here we seek for the generic solution of the alignment problem: (1) among all paths in the sequence graph, and (2) obtaining k best paths instead of just one.

2.2 Event Graph

We present the solution of *Top(k)* path problem via the special data structure called *event graph.*

First, define the expanded sequence graph $\bar{G}$ as graph G with added special EMPTY vertex with no letter on it. Connect EMPTY to all other vertices by bidirectional edges.

Each vertex (or *event*) of an event graph is a pair (E, P) of an emissive (I or M) HMM state E and a non-empty position P of $\bar{G}$ or one of two special vertices: SOURCE = (START, EMPTY) and SINK = (END, EMPTY).

A directed edge from event (E_1, P_1) to event (E_2, P_2) exists only if and only if: there is an edge from P_1 to P_2 in $\bar{G}$; emitters are consequent, i.e., there is a path in HMM from E_1 to E_2, that does not contain other emitters. Due to the used pHMM structure there could be only one such path between the states.

The weight of vertex (E, P) equals to $\log(\mathcal{E}(E, Letter(P)))$ $- \log(\mathcal{B}(Letter(P)))$. Edge weight is a sum of logarithms of the probabilities of transitions on the edge. Since these probabilities are always less or equal to 1, edge weights are always non-positive, while vertices could have positive weights due to the presence of background distribution.

One can easily see that each path from SOURCE to SINK corresponds to a global alignment of the pHMM against some path in G whereas path weight (sum of weights of edges and vertices along the path) is equal to the alignment score.

2.3 Top(1) Path Problem via Event Graph

Event graph allows one to construct the solution of Top(1) problem straight from the definition.

Algorithm 1. Top(1) path finding

Input: sequence graph G, profile HMM *pHMM*, background distribution
    Background

1 Construct event graph from G, *pHMM*, and *Background*
2 Find the best (highest scored) event path from SOURCE to SINK
3 Extract sequence path (series of positions) and the correspondent string from
  the best event path

One could easily see that the problem of finding the best event path in general does not have a finite solution. Indeed, if the event graph is acyclic or at least has no cycles of positive weight, then the solution will certainly exist. In the general case event graph could have cycles due to the presence of I-loops in pHMM and depending to a particular background distribution some of these cycles may positive weight. If there is a path from START to END that contains one of such cycles, then one can roll on it endlessly increasing the score and therefore the proper solution would not exist.

Fortunately, this problem is more theoretical than practical. In our implementation we use a simple heuristic to deal with it. We find all I-loops of positive weight in pHMM and unroll them into C (default: 10) sequential I states. This procedure is equivalent to restricting the total number of emissions. Typically there is only small fraction of such I-loops (not exceeding 2.5% for all considered HMMs and therefore the size of pHMM graph increases negligibly.

2.4   Dynamic Programming Algorithm for Best Path Extraction

For each vertex and each edge of the event graph find the best (having largest score) path from SOURCE. Denote those best score values as $DIST$(vertex) and $DIST$(edge). Note that due to absence of positive-weighted cycles those $DIST$s always exist and are finite.

The values of $DIST$s could be calculated via the Bellman-Ford algorithm [4]. However, the computational complexity of this approach is $\mathcal{O}(|V(EventGraph)| \times |E(EventGraph)|)$, that is completely impractical. Still, due to special layered event graph structure, $DIST$s could be computed significantly faster level-by-level. One possible approach to this is described in Algorithm 2.

The only non-straightforward step in Algorithm 2 is I-loop relaxation. Due to negativity of all I-loop edges we can sequentially update I-loop events $DIST$ using a priority queue as in classical Dijkstra algorithm [4]. Overall algorithm complexity is $\mathcal{O}(|pHMM| \times |E(G)| \log |V(G)|)$, where log comes from priority queue push/pop costs.

For sequence graphs $|E(G)| \sim |V(G)|$ since vertex degrees are bounded by above (e.g. by 4 for de Bruijn graphs), therefore, we will count them together denoting $|G| = |V(G)| + |E(G)|$. Thus, the overall complexity is $\mathcal{O}(|pHMM| \times |G| \log |G|)$.

Having $DIST$(SINK) calculated we can easily reconstruct the best path by backtracking: start from SINK vertex and go backwards into each vertex taking incoming edge with the highest $DIST$ until we reach SOURCE. k best paths could be found iteratively using the observation that the i-th shortest path in the sequence must branch from one of the $i-1$ shortest paths already identified (so-called Eppstein algorithm, [7]).

2.5   From Top(1) to Top(k)

Though there is a simple solution for the extraction of the best k paths from the event graph, the problem of finding top sequence graph paths is still non-trivial and could not be easily solved by extracting k best paths from SOURCE to SINK.

The first problem here is that top k alignments are not equal to top k sequences. Indeed, full event graph represents all possible alignments of the pHMM against all sequence graph paths. Each sequence path has a combinatorial number of alignments against the pHMM. And each alignment corresponds to its own path in the event graph. The highest scored path in the full event graph indeed corresponds to the best alignment of the best sequence path. However,

---

Algorithm 2. Level-by-level $DIST$ computation (optimized Bellman-Ford algorithm)

Input: Event graph (along with sequence graph G and $pHMM$ with unrolled positive I-loops)

Output: $DIST$ values for graph vertices and edges

1 $DIST(\text{SOURCE}) := 0$
2 **foreach** $edge \in Outgoing(\text{SOURCE})$ **do**
3      $DIST(edge) := weight(edge)$
4 **end**
5 **foreach** $emissive\ state\ E \in pHMM\ sorted\ in\ a\ topological\ order$ **do**
6   // Initialization from preceding levels
7   **foreach** $position\ P \in G$ **do**
8   Let vertex $v = (E, P)$
9         $DIST(v) := weight(v) + \max_e DIST(e)$, where $e \in Ingoing(v)$ and not an I-loop edge
10   **end**
11   **if** $emissive\ state\ E\ is\ I\text{-}loop\ (not\ unrolled)$ **then**
12   // I-loop relaxation
13        $Q :=$ empty priority queue
14   **foreach** $position\ P \in G$ **do**
15   Push vertex $v = (E, P)$ to Q with current $DIST(v)$ as a priority value
16   **end**
17
18   **while** $Q\ is\ not\ empty$ **do**
19   Extract vertex v with the highest priority from Q
20   **for** $edge\ e \in Outgoing\ I\text{-}loop\ edges(v)$ **do**
21   Consider vertex $v' = End(e)$
22   **if** $DIST(v') < DIST(v) + weight(e) + weight(v')$ **then**
23                 $DIST(v') := DIST(v) + weight(e) + weight(v')$
24   Increase vertex v' priority in Q
25   **end**
26   **end**
27   **end**
28   **end**
29   // Outgoing edges setup
30   **foreach** $position\ P \in G$ **do**
31   Let vertex $v = (E, P)$
32   **foreach** $edge\ e \in Outgoing(v)$ **do**
33           $DIST(e) := DIST(v) + weight(e)$
34   **end**
35   **end**
36 **end**
37 $DIST(\text{SINK})) := \max_e DIST(e)$, where edge $e \in Ingoing(\text{SINK})$

---

the second best alignment usually corresponds to the second best alignment of the same sequence path and not of some other sequence path.

The second issue is driven by particular applications. Among top k paths we would like to see really different paths rather than paths that differ only by adding/trimming a small number of letters at the beginning/end. Typically extending/trimming the sequence in such way changes the final score only slightly and variations of the best path displace all other paths from the top list.

The third issue is that the number of different sequence paths with the proper score could also be combinatorial. For the string with d single-letter variations (corresponding to a path with d bubbles) the number of different paths would be 2^d. All the paths would be slightly different and would probably have the same or almost the same score. Without additional information we could not prefer one path to another and report an answer of a reasonable (non-combinatorial) size.

This way the proper Top(k) problem statement might look like as follows: find k sufficiently (in the terms of the second issue) different paths in the initial sequence graph that maximize alignment score, but such problem seems to be very hard to solve. The only solution we know is increasing the number of extracted top alignments and further result filtering, but this approach is impractical for large k since the number of paths to extract is not known in advance. Also, the construction, storage and operations on full event graph (consisting of $|pHMM| \times |G|$ vertices) is time and memory consuming.

To deal with these issues we propose an heuristic solution that could significantly speed up Top(k) paths extraction making it possible even for very large k (say, $k > 100000$) that could be useful for complicated metagenome graphs where the number of candidate paths could be really large. This approach introduces a new structure representing alignment of the whole sequence graph against the pHMM that could be also interesting and useful by itself. Particularly, it helps to overcome the third issue, allowing one to consider relatively compact object rather than an endless number of slightly different paths.

2.6 Collapsing Event Graph

Let us consider a subgraph of the full event graph called collapsed and trimmed event graph.

Collapsed event graph is a subgraph of an event graph where for each vertex (E, P) (including SOURCE and SINK) all its ingoing and outgoing neighbours (vertices linked by ingoing and outgoing edges respectfully) have different positions. This way different START-to-END paths in collapsed event graph correspond to different paths in the initial sequence graph.

Trimmed event graph is a subgraph where each vertex does not have ingoing edge from SOURCE and ingoing edge(s) from non-SOURCE vertex at the same time. The same property is ensured for outgoing edges and SINK.

All the different SOURCE-to-SINK paths in collapsed and trimmed event graph correspond to the paths in the initial sequence graph that are not substring or superstring of each other and do not have a perfect overlap, i.e. any prefix of one path cannot not be a suffix of another.

If we perform k best paths finding algorithm on collapsed and trimmed event graph, we extract k non-trivially different sequence paths. There are multiple ways how event graph could be trimmed and collapsed. Algorithms 3 and 4 make an arbitrary subgraph of an event graph collapsed and trimmed but still keeping the best scored paths intact.

Algorithm 3. Event graph collapsing

Input: Event graph or its arbitrary subgraph

1 For each vertex and each edge calculate the $DIST$(vertex) and $DIST$(edge) values using the same DP procedure as in best path finding algorithm (Section 2.4)
2 For each vertex, group all incoming edges by position of an incident vertex. In each group find the edge with the highest score and remove other edges from the graph
3 Do steps 1 and 2 on the graph backwards: with all the edges reversed and going from SINK to SOURCE

Algorithm 4. Event graph trimming

Input: Event graph or its arbitrary subgraph

1 Compute $DIST$(edge) for each edge as in the previous algorithm
2 For each vertex incident to SOURCE: remove all incoming edges worse than the incoming edge from SOURCE. If there are still other incoming edges, remove the edge that goes from the SOURCE vertex
3 Do steps 1 and 2 on the graph with all the edges reversed considering edges going to SINK

After collapsing and trimming we perform graph cleanup by removing all vertices unreachable from SOURCE or SINK.

Algorithm Discussion and Analysis. Both algorithms preserve the best path intact since for each vertex that belongs to the best path incoming and outgoing edges with maximum $DIST$ value belong to the best path as well.

Collapsing procedure preserves existing sequence paths, but may reduce their scores. For each SOURCE-to-SINK path that got broken during the collapsing step there still exists another event path with the same corresponding sequence path. Note that collapsing removes edges, but it leaves one edge going to each position, in other words: for each sequence path $P_1, \ldots, P_n$ collapsing procedure will preserve at least one corresponding event path.

Collapsing procedure can reduce score of a path therefore yielding suboptimal alignment. If the sequence path shares prefix with another path with better score, the prefix alignment will be inherited from the other path. The same goes for the suffix. Our experiments show that this phenomenon is usually not a big problem for the real graphs. Even more, in order to address this problem one could re-align the extracted sequence paths against pHMM.

Trimming procedure could break existing sequence path but if it does so, then there does exist a path slightly different from this one with better score.

Full event graph represents all possible alignments of all graph paths against the pHMM. Collapsed and trimmed event graph represents an alignment (probably suboptimal, but still reasonable) of the whole graph against the pHMM. It could be viewed as an analogue of partial order graph alignment [10] but for graph paths rather than separate homologous sequences.

2.7   Alignment Algorithm: Collapsed Event Graph Construction

Collapsed trimmed event graph takes sufficiently less space and could be constructed on-fly, therefore we do not need to construct full graph and then collapse it. Algorithm 5 represents such an approach.

Algorithm 5. pHMM alignment against sequence graph

Input: Sequence graph G, profile HMM $pHMM$ with unrolled positive I-loops,
k — the number of best paths to be extracted

1 Start with vertex SOURCE = (START, EMPTY)
2 Iterate over pHMM emissive states in a topological order
3 **foreach** *state E* **do**
4 $\quad$ Construct new event vertices (E, P) along with ingoing edges for all positions $P \in G$
5 $\quad$ For all new edges and vertices compute $DIST$(edge) and $DIST$(vertex) and perform a step of collapsing and trimming as in Algorithms 3 and 4
6 $\quad$ Remove all vertices that definitely could not be reached from SINK
7 **end**
8 Add final vertex SINK = (END, EMPTY)
9 Perform backward collapsing and backward trimming (forward collapsing and trimming are already performed on-line)
10 Extract best k paths going from SOURCE using score values computed during backward collapsing using Eppstein algorithm [7]
11 Report best paths along with the collapsed trimming event graph

---

Two parts contribute to the algorithm complexity: collapsed event graph construction and extraction of best paths. The first part has the computational complexity of $\mathcal{O}(|pHMM| \times |G| \log |G|)$, the same as general Top(1) algorithm (see Sect. 2.4). The second part (Eppstein algorithm [7]) has complexity $\mathcal{O}(kN \log(kN))$, where N is average path length and log is from priority queue overhead. Note that in collapsed event graph incoming vertex degree is bounded like in sequence graph (usually by 4 for nucleotide graphs), therefore there is no additional term in the complexity equation.

2.8   Implementation and Heuristics

We implemented the described approximate Top(k) algorithm solution in the PATHRACER tool. PATHRACER takes as input de Bruijn graph in GFA format (only SPADES or SPADES-compatible graphs are currently supported) and nucleotide pHMM. For amino acids HMMs we perform on-fly translation that allows us to consider the initial graph as a sequence graph over extended (20 amino acids and stop codon) amino acid alphabet.

The Algorithm 5 has overall complexity $\mathcal{O}(|pHMM| \times |G| \log |G| + kN \log(kN))$, however, the first component in the sum might be intractable for large graphs. In order reduce the computation complexity we implemented several heuristics:

1. We run HMMALIGN algorithm from HMMER 3 [6] on the edges of the input sequence graph and consider only the neighborhood of suitable size of matched edges: we walk forward and backward from each matched edge not far than the length of the corresponding pHMM overhang and then join all visited positions and extract sequence subgraph. HMMER options (e.g. domain E-value thresholds) are available to control the size of seed set.
2. During the event graph construction after each step new vertices are filtered out:
   - if the continuation of alignment through this vertex will always reach a dead-end before reaching SINK vertex;
   - if the continuation of alignment through this vertex will produce stop codon before reaching SINK vertex (for amino acid HMM);
   - all vertices with the score lower than a threshold parameter (default: -250);
   - all vertices except T top score vertices. The number T is reducing while moving along the pHMM.
   This heuristic improves both the running time and memory consumption drastically (see Table 1) with no significant changes in the results.

3   Results

In order to demonstrate the viability of our solution we conducted four experiments using the variety of datasets and pHMMs.

3.1   23S Search in *E. coli str. K12* Assembly

E. coli genome contains seven ribosomal RNA (rRNA) operons. Each operon contains a 16S rRNA gene, a 23S rRNA gene, and a 5S rRNA gene (except for one operon, which contains two 5S rRNA genes) interspersed with various tRNA genes.

We considered the *E. coli str. K12* dataset from [3]. The Illumina reads were of length 100 bp with mean insert length 270 bp. SPADES 3.12 [13] assembly was performed in the normal multi-cell isolate mode with default settings and the resulting assembly graph in GFA format was obtained. Quick check of the results revealed that SPADES assembled the 16S gene on the single contig, but was unable to derive the whole sequence of 23S gene. Probably the reason is that among seven copies of 23S gene, six were inexact, still 3 variants of 23S gene were scattered over 3, 3 and 2 edges correspondingly. Other variants were absent in the assembly graph and probably were collapsed during the graph simplification procedures.

We aligned final assembly graph to prokaryotic rRNA pHMM models from [16] and all 3 variants of 23S gene were the top paths extracted by PATHRACER (see Fig. 2). No other (false) paths were produced.

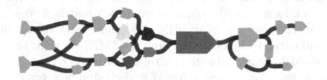

Fig. 2. *E. coli str. K12*: extracted 23S paths (yellow-violet-red, green-violet-red and blue-red) and their neighborhood (Color figure online)

3.2   16S Components of SYNTH Mock Metagenome Dataset

To showcase the ability of PATHRACER to deal with complex metagenomic data we considered 16S genes contained in SYNTH mock metagenome dataset from [17].

Synthetic community data set (SYNTH) is a set of reads from the genomic DNA mixture of 64 diverse bacterial and archaeal species (SRA acc. no. SRX200676) that was used for benchmarking the OMEGA assembler [9]. The dataset contains 109 million Illumina HiSeq 100-bp paired-end reads with mean insert size of 206 bp. The reference genomes for all 64 species forming the SYNTH dataset are known.

The assembly was performed by METASPADES 3.12 [13] with default parameters. To increase the sensitivity we considered *strain* (rather then the final *consensus*) assembly graph. We aligned the graph against 16S profile HMM from [16] and analyzed paths reported. PATHRACER reported 1088577 paths. It is a reasonable number since the corresponding graph component is very complicated (see Fig. 3) and therefore many chimeric alignments are expected.

From SILVA [15] we obtained 179 16S sequences (most of the species have several variants of this gene), constructed BLAST database from all reported paths and performed a search in this database of all 16S gene sequences from SYNTH. For 55 out of 64 species all the 16S sequences were found with >95% BLAST identity. For 22 species at least one of 16S sequences was found with 100% BLAST identity (some variations were missing).

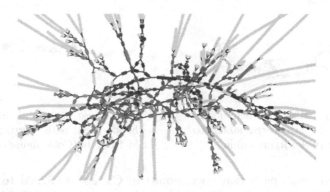

Fig. 3. SYNTH: 16S matches of known sequences and their neighborhood. Different species are colored differently (Color figure online)

3.3 Urban Wastewater Metagenome

In [12] comparative metagenomics was used to investigate the occurrence of antibiotic resistant genes in wastewater and urban surface water environments in Singapore. For this experiment we took H1 (clinical isolation ward, SRA acc. no SRR5997548) dataset and performed assembly by METASPADES 3.12 with default settings. As with SYNTH dataset we used *strain* assembly graph to increase the sensitivity of the search.

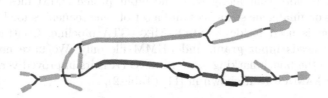

Fig. 4. bla_{CTX-M} paths and their neighborhood. Green path corresponds to CTX-M-15 family; blue and red corresponds to CTX-M-9 and CTX-M-14 respectively (Color figure online)

The assembly graph was aligned by PATHRACER against selected beta-lactamase gene pHMMs obtained from NCBIfam-AMR database [1]: bla_{CTX-M}, bla_{IMP} and bla_{TEM} among the others. The resulting paths consisting of several graph edges could be seen on Figs. 4, 5 and 6.

Different non-overlapping paths on the Fig. 4 represents several different families of bla_{CTX-M} gene: green corresponds to CTX-M-15 family, while blue and red corresponds to CTX-M-9 and CTX-M-14 respectively (differs from each other only by 2 SNPs). Note that Table S5 in [12] mentions the presence of CTX-M-15 and CTX-M-18 beta-lactamase families in this dataset, however we found that

Fig. 5. bla_{IMP} paths and their neighborhood. Red path corresponds to IMP-1 family (Color figure online)

Fig. 6. bla_{TEM} paths and their neighborhood. Red path corresponds to TEM-1 family (Color figure online)

the extracted graph path that corresponds to CTX-M-9 aligned to the reference with 100% identity. Since the amino acid sequence of the CTX-M-18 beta-lactamase differs from that of the CTX-M-9 beta-lactamase by an Ala-to-Val change at position 231 [14], the additional checking of the results of [12] might be necessary (in [12] different gene sequences were searched in reads rather than assemblies and therefore might be prone to sequencing errors).

3.4 PathRacer and MegaGTA Running Time and Memory Consumption

We compared PATHRACER performance with MEGAGTA using 16 threads on urban dataset on all 159 beta-lactamase pHMMs obtained from NCBIfam-AMR database [1]. The memory consumption and running times are presented in Table 1.

We need to note that along with the input profile HMM files MEGAGTA also requires original gene sequences and a set of gene sequences for FRAMEBOT tool [20] that is used inside XANDER/MEGAGTA pipeline. Contrary to this, PATHRACER needs input graph and pHMM file only. We have not included XANDER into the benchmarking since MEGAGTA is its improved version and it outperforms XANDER as it shown in [11] (Table 2).

Table 1. PATHRACER and MEGAGTA benchmark: running time and peak memory consumption. *urban* dataset, 159 beta-lactamase HMMs, 16 threads. PATHRACER performance with disabled event graph online filtering is also shown

Tool	Graph assembly	pHMM alignment
MEGAGTA	30 min/30 GiB	11 h/30 GiB
METASPADES + PATHRACER	1 h 50 min/20 GiB	50 min/7 GiB
METASPADES + PATHRACER (no filtering)	1 h 50 min/20 GiB	2 h 20 min/11 GiB

Table 2. PathRacer and MegaGTA results for *urban* data. blaCTX-M, blaIMP, blaTEM beta-lactamase gene families. Annotation performed by online NCBI BLAST [2], only results with 100% coverage and identity included

Gene family	As shown in [12]	MegaGTA	PathRacer
*bla*CTX-M	blaCTX-M-15; blaCTX-M-18	OXY-1-1/2/3 (177aa); CTX-M-15 (291aa); CTX-M-3 (291aa); CTX-M-9 (291aa); SFO-1 (295aa); CTX-M-14 (291aa); + 4 chimeric sequences (247aa, 203aa, 295aa, 242aa)	CTX-M-15 (291aa); CTX-M-14 (291aa); CTX-M-9 (291aa); SFO-1 (295aa)
*bla*IMP	blaIMP-1	IMP-1 precursor + 8 chimeric sequences (245aa × 8)	IMP-1 precursor
*bla*TEM	blaTEM; blaTEM-157; blaTEM-163; blaTEM-211	TEM-1 (286aa); + 43 chimeric and fragmented sequences	TEM-1 (286aa)
*bla*KPC	blaKPC2	KPC-2 (293aa); + 113 chimeric and fragmented sequences	KPC-2 (293aa)

4 Conclusion

PathRacer utilizes both the assembly graph topology and the information about the known genes encoded in profile HMM during its operation. This way the putative gene sequences could be reconstructed even from the fragmented metagenomic assemblies; it does not matter whether the gene sequence is located within the single contig or scattered across several edges of an assembly graph.

Currently PathRacer could be viewed as a standalone tool that might be integrated into analysis pipeline. Though we anticipate that deeper integration into assembler pipeline might be possible, e.g. one could utilize the information from paired-end reads to filter out more paths in the even graph, or, use pHMM alignment to supplement repeat resolution and scaffolding.

Acknowledgments. This work was supported by the Russian Science Foundation (grant 19-14-00172). The authors would like to extend a special thanks to Sergey Nurk and Tatiana Dvorkina for all the fruitful discussions that were of great help in improving the algorithms.

References

1. NCBIfam-AMR. https://ftp.ncbi.nlm.nih.gov/hmm/NCBIfam-AMR/latest/
2. Altschul, S.F., Gish, W., Miller, W., Myers, E.W., Lipman, D.J.: Basic local alignment search tool. J. Mol. Biol. **215**(3), 403–10 (1990)
3. Chitsaz, H., Yee-Greenbaum, J.L., Tesler, G., et al.: Efficient de novo assembly of single-cell bacterial genomes from short-read data sets. Nat. Biotechnol. **29**(10), 915–921 (2011)
4. Cormen, T.H., Leiserson, C.E., Rivest, R.L., Stein, C.: Introduction to Algorithms, 2nd edn. MIT Press, Cambridge (2001)
5. Eddy, S.R.: Profile hidden Markov models. Bioinformatics **14**(9), 755–763 (1998)
6. Eddy, S.R.: Accelerated profile HMM searches. PLoS Comput. Biol. **7**(10), 1–16 (2011)
7. Eppstein, D.: Finding the k shortest paths. SIAM J. Comput. **28**(2), 652–673 (1999)
8. Finn, R.D., Coggill, P., Eberhardt, R.Y., Eddy, S.R., et al.: The Pfam protein families database: towards a more sustainable future. Nucleic Acids Res. **44**(D1), D279–D285 (2016)
9. Haider, B., Ahn, T.H., Bushnell, B., Chai, J., Copeland, A., Pan, C.: Omega: an overlap-graph de novo assembler for metagenomics. Bioinformatics **30**(19), 2717–2722 (2014)
10. Lee, C., Grasso, C., Sharlow, M.F.: Multiple sequence alignment using partial order graphs. Bioinformatics **18**(3), 452–464 (2002)
11. Li, D., Huang, Y., Leung, C.M., Luo, R., Ting, H.F., Lam, T.W.: MegaGTA: a sensitive and accurate metagenomic gene-targeted assembler using iterative de Bruijn graphs. BMC Bioinform. **18**(Suppl 12), 408 (2017)
12. Ng, C., et al.: Characterization of metagenomes in urban aquatic compartments reveals high prevalence of clinically relevant antibiotic resistance genes in wastewaters. Front. Microbiol. **8**, 2200 (2017)
13. Nurk, S., Meleshko, D., Korobeynikov, A., Pevzner, P.A.: metaSPAdes: a new versatile metagenomic assembler. Genome Res. **27**(5), 824–834 (2017)
14. Poirel, L., Naas, T., Le Thomas, I., Karim, A., Bingen, E., Nordmann, P.: CTX-M-type extended-spectrum β-lactamase that hydrolyzes ceftazidime through a single amino acid substitution in the omega loop. Antimicrob. Agents Chemother. **45**(12), 3355–3361 (2001)
15. Quast, C., Pruesse, E., Yilmaz, P., Gerken, J., et al.: The SILVA ribosomal RNA gene database project: improved data processing and web-based tools. Nucleic Acids Res. **41**(D1), D590–D596 (2013)
16. Seemann, T.: Prokka: rapid prokaryotic genome annotation. Bioinformatics **30**(14), 2068–2069 (2014)
17. Shakya, M., Quince, C., Campbell, J.H., Yang, Z.K., Schadt, C.W., Podar, M.: Comparative metagenomic and rRNA microbial diversity characterization using archaeal and bacterial synthetic communities. Environ. Microbiol. **15**(6), 1882–1899 (2013)
18. Viterbi, A.: Error bounds for convolutional codes and an asymptotically optimum decoding algorithm. IEEE Trans. Inf. Theory **13**(2), 260–269 (1967)
19. Wang, Q., et al.: Xander: employing a novel method for efficient gene-targeted metagenomic assembly. Microbiome **3**, 32 (2015)
20. Wang, Q., Quensen, J.F., Fish, J.A., Kwon Lee, T., Sun, Y., et al.: Ecological patterns of nifH genes in four terrestrial climatic zones explored with targeted metagenomics using FrameBot, a new informatics tool. mBio **4**(5), e00592-13 (2013)

Genome Rearrangement, Assembly and Classification

A Uniform Theory of Adequate Subgraphs for the Genome Median, Halving, and Aliquoting Problems

Pavel Avdeyev[1]([✉]), Maria Atamanova[2], and Max A. Alekseyev[1][iD]

[1] Computational Biology Institute, The George Washington University, 45085
University Dr. Ste 305, Ashburn, VA 20147, USA
`avdeyev@gwu.edu`
[2] ITMO University, Kronverkskiy prospect 49, St. Petersburg 197101, Russia

Abstract. One of the key computational problems in comparative
genomics is the reconstruction of genomes of ancestral species based
on genomes of extant species. Since most dramatic changes in genomic
architectures are caused by genome rearrangements, this problem is often
posed as minimization of the number of genome rearrangements between
extant and ancestral genomes. The basic case of three given genomes
is known as the *genome median problem*. Whole genome duplications
(WGDs) represent yet another type of dramatic evolutionary events and
inspire the reconstruction of pre-duplicated ancestral genomes, referred
to as the *genome halving problem*. Generalization of WGDs to whole
genome multiplication events leads to the *genome aliquoting problem*.

In the present study, we generalize the adequate subgraphs approach
previously proposed for the genome median problem to the genome halving and aliquoting problems. Our study lays a theoretical foundation for
practical algorithms for the reconstruction of pre-duplicated ancestral
genomes.

Keywords: Genome rearrangement · Breakpoint graph ·
Adequate subgraph · Whole genome duplication · Genome halving ·
Genome aliquoting · Genome median

1 Introduction

Genome rearrangements are rare large-scale evolutionary events that shuffle
genomic architectures. Under the maximum parsimony assumption, the minimum number of rearrangements (known as the *rearrangement distance*) between
genomes provides a good estimate for their evolutionary remoteness. This
assumption further enables to address the ancestral genome reconstruction problem by minimizing the total distance between the ancestral and extant genomes
along the branches of the evolutionary tree. The basic case of this problem with
just three extant genomes is called the *genome median problem* (GMP), which

© Springer Nature Switzerland AG 2019
I. Holmes et al. (Eds.): AlCoB 2019, LNBI 11488, pp. 97–111, 2019.
https://doi.org/10.1007/978-3-030-18174-1_7

asks to reconstruct a single ancestral genome (*median genome*) at the minimum total distance from the given genomes.

Since genome rearrangements preserve the gene content, the GMP is considered for genomes with a uniform gene content. To account for genes present in varying number of copies in different genomes, one needs to consider other types of evolutionary events. One of the important sources of duplicated genes in genomes is the *whole genome duplication* (WGD) events, which simultaneously duplicate each chromosome of a genome. WGD events are known to happen in the evolution of yeasts [13], fishes [14], plants [12], and even mammals [8].

We distinguish between *ordinary* and *2-duplicated* genomes, where each gene is present in 1 and 2 copies, respectively. A WGD duplication of an ordinary genome R results in a *perfect* 2-duplicated genome $2R$, where not only each gene but also each gene adjacency is present in 2 copies. The *genome halving problem* (GHP) asks to reconstruct an ancestral genome R from the given descendant A of the genome $2R$ by minimizing the distance between A and $2R$. The GHP solution space is typically huge [4], making it hard to distinguish biologically relevant solutions. The *guided genome halving problem* (GGHP) improves the biological relevance by using an additional ordinary genome B and asking for an ordinary genome R that minimizes the total distance between the ordinary genomes B and R and between the 2-duplicated genome A and the perfect 2-duplicated genome $2R$. While the former distance is easy to compute (for a known genome R), computing the latter distance (called the *double distance* [17]) represents a much harder problem.

The WGD can be viewed as a particular case of a *whole genome multiplication* (WGM), which simultaneously creates $m \geq 2$ copies of each chromosome. For $m = 3$, WGM corresponds to a whole genome triplication, known to happen the evolution of eudicots. Correspondingly, the GHP and GGHP are generalized to the *genome aliquoting problem* (GAP) and the *guided genome aliquoting problem* (GGAP) [18].

Under the convenient model of *Double-Cut-and-Join* (DCJ) rearrangements [22], the GMP and GGHP are known to be NP-hard [7,17]. The GMP is the most studied problem among them, for which there exists a number of exact [5,7,19,20,23] and heuristic [5,10,15,16] algorithmic solutions. Among them the best performance is demonstrated by the software tool ASMedian, which iteratively searches for so-called *adequate subgraphs* and decomposes the GMP into smaller subproblems [15,19–21]. This strategy dramatically reduces the search space and enables ASMedian to efficiently solve the GMP for genomes with a large number of genes.

For the GGHP, there exists a number of heuristic algorithms [5,11,24,25] as well as a recent exact algorithm [5], which is based on integer linear programming and applicable only for genomes with a small number of genes. There is also a heuristic algorithm for the GGHP [11], which uses subgraphs similar to adequate but without rigorous justification.

The present study lays a theoretical foundation for the generalization of adequate subgraphs to other problems including the GAP and GGAP. Similarly

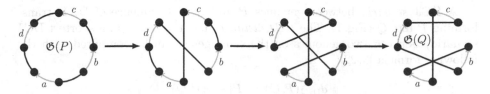

Fig. 1. A shortest DCJ scenario transforming the genome graph $\mathfrak{G}(P)$ into the genome graph $\mathfrak{G}(Q)$, where $P = (+d - c - b + a)$ and $Q = (+d + c + a - b)$ are ordinary unichromosomal circular genome.

to [19,21], our initial analysis is restricted to genomes with circular chromosomes only, while its extension to linear chromosomes will be published elsewhere. We identify all simple adequate subgraphs of small orders. For the GGHP, we found 13 adequate subgraphs of order at most 4. It turns out that 3 of these subgraphs were used in [11], and thus our results provide a justification and generalization of their approach.

2 Background

2.1 Breakpoint Graphs and DCJ Distance

Let P be an ordinary genome consisting of circular chromosomes. We represent a circular chromosome consisting of n genes as a graph cycle with n directed edges encoding genes and their strands, which alternate with n undirected edges connecting the extremities of adjacent genes. We label each directed edge with the corresponding gene x, and further label its tail and head endpoints with x^t and x^h, respectively. A collection of such cycles representing the chromosomes of P forms the *genome graph* $\mathfrak{G}(P)$ (Fig. 1). The undirected edges in $\mathfrak{G}(P)$ are called *P-edges* and form a matching in $\mathfrak{G}(P)$, called *P-matching*.

A *Double-Cut-and-Join* (DCJ) operation, also known as *2-break*, breaks a genome at two positions and glues the resulting fragments in a new order, which model common types of genome rearrangements [3,22]. A DCJ in genome P corresponds in $\mathfrak{G}(P)$ to the replacement of a pair of *P*-edges with a different pair of *P*-edges[1] on the same set of four vertices (Fig. 1).

For ordinary genomes P and Q composed of the same genes, the *breakpoint graph* $\mathfrak{G}(P,Q)$ is defined as the superposition of the genome graphs $\mathfrak{G}(P)$ and $\mathfrak{G}(Q)$ (Fig. 1). In other words, $\mathfrak{G}(P,Q)$ can be constructed by gluing the identically labeled directed edges in $\mathfrak{G}(P)$ and $\mathfrak{G}(Q)$. Similarly, the breakpoint graph $\mathfrak{G}(P_1, \ldots, P_k)$ can be defined for three or more ($k \geq 3$) genomes $P_1, \ldots, P_k$ [6]. From now on, we will ignore directed edges and assume that the breakpoint graph $\mathfrak{G}(P,Q)$ consists only of (undirected) *P*-edges and *Q*-edges, forming *P-matching* and *Q-matching*, respectively. Then the connected components of $\mathfrak{G}(P,Q)$ represent cycles formed by alternating *P*-edges and *Q*-edges, called *PQ-cycles*.

---

[1] Here we view genome P as being transformed and *P*-edges as changing.

A DCJ *scenario* between genomes P and Q is a sequence of DCJs trans-forming P into Q (Fig. 1). The *DCJ distance* (i.e., the length of a shortest DCJ scenario) between two genomes on the same genes can be computed with the following formula [3, 22]:

$$d_{\text{DCJ}}(P, Q) = |P| - c(P, Q), \tag{1}$$

where $c(P, Q)$ is the number of PQ-cycles in the breakpoint graph $\mathfrak{G}(P, Q)$ and $|P|$ is the number of genes in P.

Let us formulate the genome median problem under the DCJ model, which we generalize to an arbitrary number of genomes:[2]

Genome Median Problem (GMP). *For given ordinary genomes $B_1, \ldots, B_n$ on the same genes, find an ordinary genome R minimizing the total distance $\sum_{i=1}^{n} d_{\text{DCJ}}(B_i, R)$.*

2.2 Contracted Breakpoint Graphs and Aliquoting Problems

We adopt the model of WGDs, where a WGD simultaneously duplicates each circular chromosome into either a single 2-duplicated circular chromosome or two identical circular chromosomes [1]. Similarly, we assume that a WGM can either create $m \geq 2$ identical copies of a circular chromosome or possibly combine some of these copies into single chromosomes. Hence, a WGM of an ordinary genome R can multiply each circular chromosome of R in $p(m)$ ways, where $p(m)$ is the number of integer partitions of m (with $p(2) = 2$, $p(3) = 3$, $p(4) = 5$, ...). Let $\Omega_m(R)$ be the set of *perfect m-duplicated genomes mR* (i.e., genomes immediately resulted from a WGM). It follows that $|\Omega_m(R)| = p(m)^{\text{cchr}(R)}$, where $\text{cchr}(R)$ is the number of circular chromosomes in R.

Let A be an m-duplicated genome. We remark that in the genome graph $\mathfrak{G}(A)$, the directed edges appear in m identically labeled copies (Fig. 2a). By gluing all directed edges with identical labels into single edges, we obtain the *contracted genome graph* $\hat{\mathfrak{G}}(A)$ (Fig. 2b). We define *A-components* in $\hat{\mathfrak{G}}(A)$ as the connected components formed by A-edges. We remark that in the case $m = 2$, A-components form cycles (called *A-cycles*).

For an ordinary genome R and integer $m \geq 2$, it can be easily seen that the contracted genome graph $\hat{\mathfrak{G}}(mR)$ does not depend on a particular choice of $mR \in \Omega_m(R)$, and its R-edges form multi-edges composed of m parallel edges each. We refer to such multi-edges as *mR-edges*. It is clear that the mR-edges form a matching in $\hat{\mathfrak{G}}(mR)$. Replacing each mR-edge with an R-edge in $\hat{\mathfrak{G}}(mR)$ transforms it into the genome graph $\mathfrak{G}(R)$.

For an m-duplicated genome A and an ordinary genome R composed of the same genes (present in m and 1 copies in A and R, respectively), the *contracted breakpoint graph* $\hat{\mathfrak{G}}(A, R)$ is defined as the superposition of $\hat{\mathfrak{G}}(A)$ and $\mathfrak{G}(R)$, and can be constructed in the same way as breakpoint graphs (Fig. 2b, c, d). The

---

[2] The classic formulation of the GMP corresponds to $n = 3$.

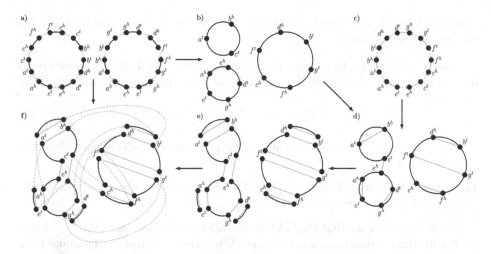

Fig. 2. For a 2-duplicated genome $A = (-a-b+g+d+f+g+e)(-a+c-f-c-b-d-e)$ and an ordinary genome $R = (-a-b-d-g+f-c-e)$, (a) the genome graph $\mathfrak{G}(A)$; (b) the contracted genome graph $\hat{\mathfrak{G}}(A)$; (c) the genome graph $\mathfrak{G}(R)$; (d) the contracted breakpoint graph $\hat{\mathfrak{G}}(A, R)$; (e) a maximal AR-cycle decomposition C of $\hat{\mathfrak{G}}(A, 2R)$; and (f) the breakpoint graph $\mathfrak{G}(A, X)$ having the same cycle structure as C, for some genome $X \in \Omega_2(R)$ and some labeling of gene copies of A and X.

A-edges and the R-edges in $\hat{\mathfrak{G}}(A, R)$ form A-components (A-cycles if $m = 2$) and R-matching, respectively. Similarly, the *contracted breakpoint graph* $\hat{\mathfrak{G}}(A, mR)$ of an m-duplicated genome A and a perfect m-duplicated genome mR is defined as the superposition of the contracted genome graphs $\hat{\mathfrak{G}}(A)$ and $\hat{\mathfrak{G}}(mR)$. The graph $\hat{\mathfrak{G}}(A, mR)$ can also be obtained from $\hat{\mathfrak{G}}(A, R)$ by replacing each R-edge with an mR-edge.

Since multiplication of a circular chromosome can happen in $p(m)$ ways, the DCJ distance between an *m-duplicated genome* and a perfect m-duplicated genome (called the *aliquoting DCJ distance*) can be computed with the following formula [5]:

$$d_{\mathrm{DCJ}}^m(A, R) = \min_{mR \in \Omega_m(R)} d_{\mathrm{DCJ}}(A, mR).$$

Since each vertex in $\hat{\mathfrak{G}}(A, mR)$ is incident to m A-edges and m R-edges, $\hat{\mathfrak{G}}(A, mR)$ can be decomposed into a collection of AR-cycles. We refer to the graph composed of these cycles (each forming its own connected component) as a *decomposition* of $\hat{\mathfrak{G}}(A, mR)$ (Fig. 2e). For a decomposition D, we define $c_D(A, R)$ as the number of AR-cycles in D. Let $\mathcal{D}(A, R)$ be the set of all decompositions of $\hat{\mathfrak{G}}(A, mR)$. The following theorem establishes a link between the aliquoting DCJ distance and the decompositions of $\hat{\mathfrak{G}}(A, mR)$:

Theorem 1 ([2,5]). *Let A be an m-duplicated genome and R be an ordinary genome composed of the same genes. For any decomposition D of $\hat{\mathfrak{G}}(A, mR)$, there exists a genome $X \in \Omega_m(R)$ and a labeling of gene copies in A and X*

(turning A and X into ordinary genomes with mn genes) such that $\mathfrak{G}(A, X)$ has the same structure as D.

Theorem 1 implies that computing the aliquoting DCJ distance reduces to finding an *optimal* decomposition of $\overset{\ast}{\mathfrak{G}}(A, mR)$ that maximizes $c_D(A, R)$. That is,

$$d_{\mathrm{DCJ}}^m(A, R) = m|R| - \max_{D \in \mathcal{D}(A, R)} c_D(A, R). \qquad (2)$$

Now, let us recall the genome aliquoting problem [18], which generalizes the genome halving problem:

Genome Aliquoting Problem (GAP). *For a given m-duplicated genome A, find an ordinary genome R minimizing $d_{\mathrm{DCJ}}^m(A, R)$.*

From (2), it follows that the GAP is equivalent to finding an ordinary genome R that maximizes $\max_{D \in \mathcal{D}(A,R)} c_D(A, R)$. While the GAP has a polynomial-time solution for $m = 2$ [1,9], its computational complexity for $m \geq 3$ is unknown.

The GAP solution space is typically huge [4], which makes it hard to distinguish biologically relevant solutions. This issue can be addressed by taking into account additional ordinary genomes and posing the following problem:

Guided Genome Aliquoting Problem (GGAP). *Given an m-duplicated genome A and ordinary genomes $B_1, \ldots, B_n$ on the same genes, find an ordinary genome R minimizing the total distance to genomes $A, B_1, \ldots, B_n$, i.e., $d_{\mathrm{DCJ}}^m(A, R) + \sum_{i=1}^n d_{\mathrm{DCJ}}(R, B_i)$.*

The guided genome halving problem (GGHP) [17] represents a particular case of the GGAP with $m = 2$, $n = 1$.

3 Methods

3.1 Family of Alternating Graphs

We start with defining an abstraction of breakpoint graphs and contracted breakpoint graphs. Let $V(G)$ and $E(G)$ be the set of vertices and the (multi)set of edges of a (multi)graph G, respectively. For a subset of vertices $U \subset V(G)$, we denote by G_U the induced subgraph of G on the vertex set U, and $U^c := V(G) \backslash U$.

Alternating Graphs. We call a graph cycle or path *alternating* if its edges are colored into two distinct colors and alternate between them (in particular, each alternating cycle has an even length).

Let P and Q be distinct colors. We say that a graph cycle C is a *labeled PQ-colored cycle* if each vertex in C has a label, each edge in C is colored into P or Q, and C is alternating. The labels, say x and y, at the endpoints of each edge of C induce the multiset $\{x, y\}$ as a label of this edge.

For positive integer m, a vertex-disjoint union of labeled PQ-cycles $\mathcal{C} = C_1 \sqcup \cdots \sqcup C_n$ forms a *simple (P^m, Q^m)-alternating graph* if

(i) every vertex label in $\mathcal{C}$ is present in m copies (in other words, each element of the multiset of the vertex labels in $\mathcal{C}$ has multiplicity m);

(ii) every edge label $\{x, y\}$ of a Q-colored edge appears as such in $\mathcal{C}$ exactly m times (in other words, each element of the multiset of the Q-colored edge labels in $\mathcal{C}$ has multiplicity m).

For a simple (P^m, Q^m)-alternating graph G, we denote by $c(G)$ and $L(G)$ the number of PQ-colored cycles and the set of vertex labels in G, respectively. We will omit the upper index m when $m = 1$.

It is easy to see that the breakpoint graph of two ordinary genomes P and Q forms a simple (P, Q)-alternating graph as well as a simple (Q, P)-alternating graph. More generally, by Theorem 1, any decomposition of the contracted breakpoint graph for an m-duplicated genome A and a perfect m-duplicated genome mR forms a simple (A^m, R^m)-alternating graph, but not a simple (R^m, A^m)-alternating graph unless A is also perfect. However, in general we do not require the vertex labels in simple (P^m, Q^m)-alternating graphs to correspond to gene extremities, and therefore we can consider such graphs outside the context of genomes.[3]

Similarly to breakpoint graphs, the simple alternating graphs can be generalized to more than two colors as follows. Let n be a positive integer and $B_1, B_2, \ldots, B_n, R$ be distinct colors. For positive integers $m_1, m_2, \ldots, m_n$, and $q = m_1 + \cdots + m_n$, a $(B_1^{m_1}, \ldots, B_n^{m_n}, R^q)$-*alternating graph* G is the vertex-disjoint union of graphs $G_1 \sqcup \cdots \sqcup G_n$, where G_i is a simple $(B_i^{m_i}, R^{m_i})$-alternating graph, and every edge label $\{x, y\}$ of the R-colored edge appears as such in G exactly q times. It follows that $L(G_1) = L(G_2) = \cdots = L(G_n)$, and we define $L(G) := L(G_1)$. Moreover, for any subset $J \subseteq L(G)$, we define $V(J) \subseteq V(G)$ as the subset of vertices with labels from J. We further define $c(G) := \sum_{i=1}^{n} c(G_i)$. We will also refer to G as an *alternating graph* if its parameters are clear from the context.

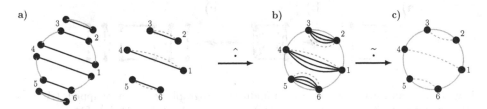

Fig. 3. (a) An (A^2, B, R^3)-alternating graph G with $L(G) = \{1, 2, 3, 4, 5, 6\}$; (b) the contracted (A^3, B, R^3)-alternating graph $\hat{G}$; and (c) the stripped (A^3, B)-alternating graph $\bar{G} = \hat{\bar{G}}$, where A, B, and R are *solid gray*, *dashed gray*, and *solid black* colors, respectively.

---

[3] This enables us to use transformations of alternating graphs, where intermediate graphs do not have genomic interpretations.

Contracted Alternating Graphs. Let n be a positive integer and $B_1, \ldots, B_n, R$ be distinct colors. For positive integers $m_1, m_2, \ldots, m_n$, and $q = m_1 + \cdots + m_n$, a *contracted* $(B_1^{m_1}, \ldots, B_n^{m_n}, R^q)$-*alternating graph* Y is a graph such that

(i)   each edge of Y is colored into a color from $\{B_1, \ldots, B_n, R\}$;
(ii) each vertex of Y has a unique label and is incident to m_1 B_1-colored edges, m_2 B_2-colored edges, ..., m_n B_n-colored edges, and q R-colored edges;
(iii) the R-colored edges of Y form multi-edges composed of q parallel edges each.

It follows that R-colored multi-edges form a perfect matching in Y (Fig. 3b).

Since each vertex in Y is incident to q R-colored edges and m_i B_i-colored edges for each $i \in [n]$, Y can be decomposed into a collection of graphs $\{G_1, \ldots, G_n\}$, where each G_i is a simple $(B_i^{m_i}, R^{m_i})$-alternating graph. In other words, Y can be decomposed into an $(B_1^{m_1}, \ldots, B_n^{m_n}, R^q)$-alternating graph $G_1 \sqcup \cdots \sqcup G_n$. We denote by $\mathfrak{D}(Y)$ the set of all possible alternating graphs forming decompositions of Y. Let $c_{\max}(Y)$ be the maximum number of cycles in such an alternating graph, i.e.,

$$c_{\max}(Y) := \max_{G \in \mathfrak{D}(Y)} c(G). \tag{3}$$

More generally, for any induced subgraph H of Y and any $G \in \mathfrak{D}(Y)$, the induced subgraph $G_{V(L(H))}$ forms a decomposition of H into alternating cycles and paths (Fig. 4b, c). Let $\mathfrak{D}(H)$ be the set of all such decompositions (notice that this definition of $\mathfrak{D}(H)$ for $H = Y$ is consistent with the earlier definition of $\mathfrak{D}(Y)$). Among the decompositions in $\mathfrak{D}(H)$, we distinguish *maximal decompositions* with the maximum possible number of alternating cycles, denoted $c_{\max}(H)$ (Fig. 4c). We remark that the definition of $c_{\max}(H)$ is consistent with the earlier definition of $c_{\max}(Y)$ (i.e., $c_{\max}(H) = c_{\max}(Y)$ if $H = Y$).

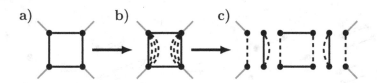

Fig. 4. A transformation of (a) an induced subgraph Z' of a stripped (A^2, B)-alternating graph (with $\max_{H \in \mathfrak{A}(Z')} c(H) = 3$) into (b) a subgraph Y' of a contracted (A^2, B, R^3)-alternating graph, and then into (c) a subgraph G' of an (A^2, B, R^3)-alternating graph with $c(G') = 3$. Here edges of colors A, B, R are shown as black, gray, and dashed, respectively.

Let G be a $(B_1^{m_1}, \ldots, B_n^{m_n}, R^q)$-alternating graph. For any subgraph H of G, we define $\hat{H}$ as the result of gluing vertices of H with the same labels together. It is clear that $\hat{G}$ forms a contracted $(B_1^{m_1}, \ldots, B_n^{m_n}, R^q)$-alternating graph. Moreover, it follows that for any $G \in \mathfrak{D}(Y)$, $\hat{G} = Y$.

Stripped Alternating Graphs. Let n be a positive integer and $B_1, \ldots, B_n$ be distinct colors. For positive integers $m_1, m_2, \ldots, m_n$, a *stripped* $(B_1^{m_1}, \ldots, B_n^{m_n})$-*alternating graph* (or simply a *stripped alternating graph* if the parameters are clear from the context) Z is a graph such that

(i) each edge of Z is colored into a color from $\{B_1, \ldots, B_n\}$;

(ii) each vertex of Z has a unique label and is incident to m_1 B_1-colored edges, m_2 B_2-colored edges, $\ldots$, m_n B_n-colored edges.

For a given color R different from $B_1, \ldots, B_n$ and an element of $\mathcal{M}(Z)$, adding R-colored multi-edges of multiplicity $q = m_1 + \cdots + m_n$ connecting vertices in the pairs to Z results in a contracted $(B_1^{m_1}, \ldots, B_n^{m_n}, R^q)$-alternating graph. Let $\mathfrak{C}(Z)$ be the set of all such contracted alternating graphs, and $\mathfrak{A}(Z) := \bigcup_{Y \in \mathfrak{C}(Z)} \mathfrak{D}(Y)$.

For any induced subgraph H of Z with an even number of vertices, we define $\mathcal{M}(H)$ as the set of all partitions of $V(H)$ into disjoint pairs. For a given color R different from $B_1, \ldots, B_n$ and an element of $\mathcal{M}(H)$, adding R-colored multi-edges of multiplicity $q = m_1 + \cdots + m_n$ connecting vertices in the pairs to H results in an induced subgraph of some contracted $(B_1^{m_1}, \ldots, B_n^{m_n}, R^q)$-alternating graph (Fig. 4a, b). Let $\mathfrak{C}(H)$ be the set of all such subgraphs and $\mathfrak{A}(H) := \bigcup_{W \in \mathfrak{C}(H)} \mathfrak{D}(W)$. If $H = Z$, then the resulting graph forms a contracted $(B_1^{m_1}, \ldots, B_n^{m_n}, R^q)$-alternating graph, $\mathfrak{C}(Z)$ is the set of all such contracted alternating graphs, and elements of $\mathfrak{A}(Z)$ form $(B_1^{m_1}, \ldots, B_n^{m_n}, R^q)$-alternating graphs.

Let Y be a contracted $(B_1^{m_1}, \ldots, B_n^{m_n}, R^q)$-alternating graph. For any induced subgraph H of Y, we define $\tilde{H}$ as the result of removing the R-colored edges from H. It is clear that $\tilde{Y}$ is a stripped $(B_1^{m_1}, \ldots, B_n^{m_n})$-alternating graph. Moreover, it follows that for any $Y \in \mathfrak{C}(Z)$, $\tilde{Y} = Z$.

Let G be a $(B_1^{m_1}, \ldots, B_n^{m_n}, R^q)$-alternating graph. For any induced subgraph H of G, we define $\bar{H} := \tilde{\hat{H}}$, where $\hat{H}$ is stripped $(B_1^{m_1}, \ldots, B_n^{m_n}, R^q)$-alternating graph. It follows that for any $G \in \mathfrak{A}(Z)$, $\bar{G} = Z$.

3.2   Alternating Graph Decomposition Problem

We pose the following problem that generalizes the GMP, GAP, GGAP.

Alternating Graph Decomposition Problem (AGDP). *For a given stripped alternating graph Z, find $G = \mathrm{argmax}_{G' \in \mathfrak{A}(Z)} c(G')$ (such G is called optimal).*

Theorem 2.

(i) *For an integer $n \geq 3$, the GMP for ordinary genomes $B_1, \ldots, B_n$ on the same genes is equivalent to the AGDP for the breakpoint graph $\mathfrak{G}(B_1, \ldots, B_n)$;*

(ii) *For an integer $m \geq 3$, the GAP for an m-duplicated genome A is equivalent to the AGDP for the contracted genome graph $\hat{\mathfrak{G}}(A)$;*

(iii) For integers $m \geq 2$ and $n \geq 1$, the GGAP for ordinary genomes $B_1, \ldots, B_n$ and an m-duplicated genome A on the same genes is equivalent to the AGDP for the contracted breakpoint graph $\hat{\mathfrak{G}}(A, B_1, \ldots, B_n)$.

Proof. (i) For given ordinary genomes $B_1, \ldots, B_n$ on the same l genes, the breakpoint graph $Z = \mathfrak{G}(B_1, \ldots, B_n)$ forms a stripped $(B_1, \ldots, B_n)$-alternating graph. Recall that the GMP asks for an ordinary genome R that minimizes the total distance $\sum_{i=1}^{n} d_{\text{DCJ}}(R, B_i)$. By formula (1), we have

$$\sum_{i=1}^{n} d_{\text{DCJ}}(R, B_i) = n \cdot l - \sum_{i=1}^{n} c(R, B_i).$$

Hence, the GMP asks for an ordinary genome R (viewed as an element of $\mathcal{M}(Z)$) such that

$$R = \operatorname*{argmax}_{R' \in \mathcal{M}(Z)} \sum_{i=1}^{n} c(R', B_i).$$

In other words, R corresponds to $\operatorname{argmax}_{Y \in \mathfrak{C}(Z)} c_{\max}(Y)$. Since for each $i \in [n]$, $\mathfrak{G}(B_i, R)$ is a simple alternating graph, we have $|\mathfrak{D}(Z)| = 1$. By (3), $c_{\max}(Y) = \sum_{i=1}^{n} c(R, B_i)$. Thus, the GMP for $B_1, \ldots, B_n$ is equivalent to the AGDP for Z.

(ii) For a given m-duplicated genome A on l unique genes, the contracted genome graph $Z = \hat{\mathfrak{G}}(A)$ forms a stripped (A^m)-alternating graph. Recall that the GAP asks for an ordinary genome R that minimizes $d_{\text{DCJ}}^m(A, R)$. By formula (2), we have

$$d_{\text{DCJ}}^m(A, R) = m \cdot l - \max_{D \in \mathcal{D}(A,R)} c_D(A, R).$$

Hence, the GAP asks for an ordinary genome R (viewed as an element of $\mathcal{M}(Z)$) such that

$$R = \operatorname*{argmax}_{R' \in \mathcal{M}(Z)} \max_{D \in \mathcal{D}(A,R')} c_D(A, R').$$

In other words, R corresponds to $\operatorname{argmax}_{Y \in \mathfrak{C}(Z)} c_{\max}(Y)$. Since by Theorem 1, any decomposition is a simple (A^m, R^m)-alternating graph, elements from $\mathfrak{D}(Y)$ correspond to elements from $\mathcal{D}(A, R)$. By (3), $c_{\max}(Y) = \max_{G \in \mathfrak{D}(Y)} c(G) = \max_{D \in \mathcal{D}(A,R)} c_D(A, R)$. Thus, the GAP for A is equivalent to the AGDP for Z.

(iii) For a given m-duplicated genome A and ordinary genomes $B_1, \ldots, B_n$ on the same l unique genes, the contracted breakpoint graph $Z = \hat{\mathfrak{G}}(A, B_1, \ldots, B_n)$ forms a stripped $(A^m, B_1, \ldots, B_n)$-alternating graph (Fig. 3c). Recall that the GGAP asks for an ordinary genome R that minimizes the total distance $d_{\text{DCJ}}^m(A, R) + \sum_{i=1}^{n} d_{\text{DCJ}}(R, B_i)$. By formulas (1) and (2), we have

$$d_{\text{DCJ}}^m(A, R) + \sum_{i=1}^{n} d_{\text{DCJ}}(R, B_i) = (m+n) \cdot l - \max_{D \in \mathcal{D}(A,R)} c_D(A, R) - \sum_{i=1}^{n} c(R, B_i).$$

Hence, the GGAP asks for an ordinary genome R (viewed as an element of $\mathcal{M}(Z)$) such that

$$R = \operatorname*{argmax}_{R'} \left(\sum_{i=1}^{n} c(R', B_i) + \max_{D \in \mathcal{D}(A,R')} c_D(A, R') \right).$$

Again, R corresponds to $\mathrm{argmax}_{Y \in \mathfrak{C}(Z)}\, c_{\max}(Y)$. Similarly to (1) and (2), the elements of $\mathfrak{D}(Y)$ correspond to the elements of $\mathcal{D}(A, R)$ that have simple (B_i, R)-alternating graphs for each $i \in [n]$. Since by (3) we have $c_{\max}(Y) = \sum_{i=1}^n c(R, B_i) + \max_{D \in \mathcal{D}(A,R)} c_D(A, R)$, the GGAP for $A, B_1, \ldots, B_n$ is equivalent to the AGDP for Z. $\qquad\square$

Let Z be a stripped alternating graph and U be a proper subset $V(Z)$ of even size. If there exist optimal $H' \in \mathfrak{A}(Z_U)$ and $H'' \in \mathfrak{A}(Z_{U^c})$ such that the union $H' \cup H'' \in \mathfrak{A}(Z)$ is optimal, then one can reduce finding an optimal element in $\mathfrak{A}(Z)$ to finding those independently in $\mathfrak{A}(Z_U)$ and $\mathfrak{A}(Z_{U^c})$. In the next section we provide a characterization of $U \subsetneq V(Z)$, for which there exist such H' and H''. The ability to identify such vertex subsets U in the given graph Z enables design of a divide-and-conquer algorithm for the AGDP (to be described elsewhere).

3.3 Decomposers and Adequate Subgraphs

Let Y be a contracted $(B_1^{m_1}, \ldots, B_n^{m_n}, R^q)$-alternating graph, and $U \subsetneq V(Y)$. We call an R-colored edge $\{u, v\}$ in Y *U-crossing* if $|\{u, v\} \cap U| = 1$. Correspondingly, the graph Y is *U-crossing* if it has at least one U-crossing edge.

Definition 3. *For a stripped alternating graph Z and $U \subsetneq V(Z)$, the subgraph Z_U is called*

- a strong decomposer *if for any* optimal alternating graph $G \in \mathfrak{A}(Z)$, $\hat{G}$ *is not U-crossing;*
- a decomposer *if for some* optimal alternating graph $G \in \mathfrak{A}(Z)$, $\hat{G}$ *is not U-crossing.*

It easy to see that if the subgraph Z_U is a (strong) decomposer, then there exist $H' \in \mathfrak{A}(Z_U)$ and $H'' \in \mathfrak{A}(Z_{U^c})$ such that $H' \cup H'' \in \mathfrak{A}(Z)$ is optimal. However, while Definition 3 describes induced subgraphs that can be used for dividing the AGDP into subproblems, it remains unclear how to identify such subgraphs. The following definition and theorem establish an important class of such subgraphs, which can be efficiently identified.

Definition 4. *Let Z be a stripped $(B_1^{m_1}, \ldots, B_n^{m_n})$-alternating graph, and $U \subsetneq V(Z)$ such that $|U|$ is even.*

- *If $\max_{H \in \mathfrak{A}(Z_U)} c(H) \geq \frac{|U|}{4} \sum_{i=1}^n m_i$, then Z_U is called* adequate subgraph;
- *If $\max_{H \in \mathfrak{A}(Z_U)} c(H) > \frac{|U|}{4} \sum_{i=1}^n m_i$, then Z_U is called* strongly adequate subgraph.

A (strongly) adequate subgraph Z_U is called simple *if for any $W \subsetneq U$, Z_W is not (strongly) adequate.*

Theorem 5. *Let Z be a stripped alternating graph, and $U \subsetneq V(Z)$ such that $|U|$ is even. If Z_U is (strongly) adequate, then Z_U is a (strong) decomposer.*

The proof of Theorem 5 will be published elsewhere in a full-length paper.

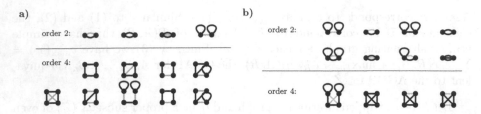

Fig. 5. Simple adequate subgraphs of order 2 and 4 that may appear in (a) stripped (A^2, B)-alternating graphs and (b) stripped (A^3, B)-alternating graphs, where A is a *black* color and B is a *gray* color.

4   Results

Adequate subgraphs were initially introduced in a branch-and-bound algorithm for the GMP for three ordinary genomes [19]. The algorithm relies on simple adequate subgraphs of small orders (listed in [21, Fig. 7]), searches them in the breakpoint graph of the given genomes, and uses the known subsolutions to reduce the problem size. It is shown that this approach works quite well for many real GMP instances.

In the present study, we generalize the theory of adequate subgraphs to other ancestral genome reconstruction problems. In particular, the definition of (strongly) adequate subgraphs in [21] represents a particular case of our Definition 4 with $\mathfrak{A}(Z) = \mathfrak{C}(Z)$. Furthermore, Theorem 2 in [21] is a particular case of our Theorem 5.

While there exist simple adequate subgraphs of an arbitrarily large order [19], the order of detected adequate subgraphs represents a trade-off between the complexity of detecting adequate subgraphs and that of the brute-force solution. Indeed, simple adequate subgraphs of small orders occur with a much higher probability than those of larger orders in random graphs [19]. Furthermore, since the number of different simple adequate subgraphs grows dramatically as the order increases, the complexity of their detection in a given graph also increases accordingly. We therefore focus on enumeration of the simple adequate subgraphs of only small orders for the GGHP (addressing reconstruction of ancestral pre-duplicated genomes) as well as for the GAP and GGAP for $m = 3$ (addressing reconstruction of ancestral pre-triplicated genomes).

Guided Genome Halving Problem. By Theorem 2, the GGHP corresponds to the AGDP for the contracted breakpoint graph for a given 2-duplicate genome A and ordinary genome B. We identified all simple adequate subgraphs of order 2 and 4 of a stripped (A^2, B)-alternating graph (Fig. 5a).

Genome Aliquoting Problem. By Theorem 2, the GAP for a given 3-duplicate genome A corresponds to the AGDP for the contracted genome graph of A. We identified all simple adequate subgraphs of order 2, 4, and 8 of a stripped (A^3)-alternating graph (Fig. 6).

Guided Genome Aliquoting Problem. By Theorem 2, the GGAP with $m = 3$ corresponds to the AGDP for the contracted breakpoint graph for a given 3-duplicate genome A and ordinary genome B. We identified all simple adequate subgraphs of order 2, 4 of order 2, 4 of a stripped (A^3, B)-alternating graph (Fig. 5b).

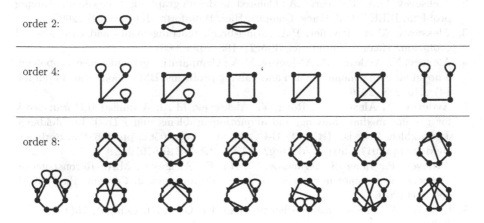

order 2:

order 4:

order 8:

Fig. 6. Simple adequate subgraphs of orders 2, 4, and 8 that may appear in a stripped (A^3)-alternating graphs, where A is a *black* color.

5   Discussion

Among the existing approaches for the genome median problem, the best performance is demonstrated by the one based on searching adequate subgraphs [15, 19–21]. The present study lays theoretical foundation for the generalization of adequate subgraphs to the genome aliquoting and halving problems. We remark that the main result of [21] follows from this extension as a corollary.

Using the generalized theory, we identified and reported all simple adequate subgraphs of small order. In particular, for the guided genome halving problem we discovered 13 adequate subgraphs of small order, out of which only 3 were earlier reported without a justification in [11]. We further reported adequate subgraphs of small order for aliquoting of triplicated genomes, which are applicable for analysis of whole genome triplication events in the plan evolution.

In future work, we will use our theoretical framework to develop branch-and-bound and heuristic algorithms for the GGHP, GGAP, and GAP (similarly to those for the GMP proposed in [15, 21]) and extend them to genomes with linear chromosomes. We also plan to release a user-friendly software tool that implements these algorithms.

Acknowledgements. The work of Maria Atamanova was supported by the Government of the Russian Federation (Grant 08-08) and JetBrains Research.

References

1. Alekseyev, M.A., Pevzner, P.A.: Whole genome duplications, multi-break rearrangements, and genome halving problem. In: Proceedings of the Eighteenth Annual ACM-SIAM Symposium on Discrete Algorithms (SODA 2007), pp. 665–679. Society for Industrial and Applied Mathematics, Philadelphia (2007)
2. Alekseyev, M.A., Pevzner, P.A.: Colored de Bruijn graphs and the genome halving problem. IEEE/ACM Trans. Comput. Biol. Bioinform. **4**(1), 98–107 (2007)
3. Alekseyev, M.A., Pevzner, P.A.: Multi-break rearrangements and chromosomal evolution. Theor. Comput. Sci. **395**(2), 193–202 (2008)
4. Alexeev, N., Avdeyev, P., Alekseyev, M.A.: Comparative genomics meets topology: a novel view on genome median and halving problems. BMC Bioinform. **17**(Suppl. 14), 213–223 (2016)
5. Avdeyev, P., Alexeev, N., Rong, Y., Alekseyev, M.A.: A unified ILP framework for genome median, halving, and aliquoting problems under DCJ. In: Meidanis, J., Nakhleh, L. (eds.) (RECOMB-CG). LNCS, vol. 10562, pp. 156–178. Springer, Heidelberg (2017). https://doi.org/10.1007/978-3-319-67979-2_9
6. Avdeyev, P., Jiang, S., Aganezov, S., Hu, F., Alekseyev, M.A.: Reconstruction of ancestral genomes in presence of gene gain and loss. J. Comput. Biol. **23**(3), 150–164 (2016)
7. Caprara, A.: The reversal median problem. INFORMS J. Comput. **15**(1), 93–113 (2003)
8. Dehal, P., Boore, J.L.: Two rounds of whole genome duplication in the ancestral vertebrate. PLoS Biol. **3**(10), e314 (2005)
9. El-Mabrouk, N., Sankoff, D.: The reconstruction of doubled genomes. SIAM J. Comput. **32**(3), 754–792 (2003)
10. Gao, N., Yang, N., Tang, J.: Ancestral genome inference using a genetic algorithm approach. PLOS ONE **8**(5), 1–6 (2013)
11. Gavranović, H., Tannier, E.: Guided genome halving: provably optimal solutions provide good insights into the preduplication ancestral genome of saccharomyces cerevisiae. Pac. Symp. Biocomput. **15**, 21–30 (2010)
12. Guyot, R., Keller, B.: Ancestral genome duplication in rice. Genome **47**(3), 610–614 (2004)
13. Kellis, M., Birren, B.W., Lander, E.S.: Proof and evolutionary analysis of ancient genome duplication in the yeast Saccharomyces cerevisiae. Nature **428**(6983), 617–624 (2004)
14. Postlethwait, J.H., et al.: Vertebrate genome evolution and the zebrafish gene map. Nat. Genet. **18**(4), 345–349 (1998)
15. Rajan, V., Xu, A.W., Lin, Y., Swenson, K.M., Moret, B.M.E.: Heuristics for the inversion median problem. BMC Bioinform. **11**(Suppl. 1), S30 (2010)
16. Ruofan, X., Yu, L., Jun, Z., Bing, F., Jijun, T.: A median solver and phylogenetic inference based on double-cut-and-join sorting. J. Comput. Biol. **25**(3), 302–312 (2018)
17. Tannier, E., Zheng, C., Sankoff, D.: Multichromosomal median and halving problems under different genomic distances. BMC Bioinform. **10**, 120 (2009)
18. Warren, R., Sankoff, D.: Genome aliquoting with double cut and join. BMC Bioinform. **10**(Suppl. 1), S2 (2009)
19. Xu, A.W.: A fast and exact algorithm for the median of three problem: a graph decomposition approach. J. Comput. Biol. **16**(10), 1369–1381 (2009)

20. Xu, A.W.: DCJ median problems on linear multichromosomal genomes: graph representation and fast exact solutions. In: Ciccarelli, F.D., Miklós, I. (eds.) RECOMB-CG 2009. LNCS, vol. 5817, pp. 70–83. Springer, Heidelberg (2009). https://doi.org/10.1007/978-3-642-04744-2_7
21. Xu, A.W., Sankoff, D.: Decompositions of multiple breakpoint graphs and rapid exact solutions to the median problem. In: Crandall, K.A., Lagergren, J. (eds.) WABI 2008. LNCS, vol. 5251, pp. 25–37. Springer, Heidelberg (2008). https://doi.org/10.1007/978-3-540-87361-7_3
22. Yancopoulos, S., Attie, O., Friedberg, R.: Efficient sorting of genomic permutations by translocation, inversion and block interchange. Bioinformatics **21**(16), 3340–3346 (2005)
23. Zhang, M., Arndt, W., Tang, J.: An exact solver for the DCJ median problem. In: Pacific Symposium on Biocomputing 2009, pp. 138–149. World Scientific (2009)
24. Zheng, C., Zhu, Q., Adam, Z., Sankoff, D.: Guided genome halving: hardness, heuristics and the history of the Hemiascomycetes. Bioinformatics **24**(13), i96 (2008)
25. Zheng, C., Zhu, Q., Sankoff, D.: Genome halving with an outgroup. Evol. Bioinform. **2**, 295–302 (2006)

Lightweight Metagenomic Classification via eBWT

Veronica Guerrini[iD] and Giovanna Rosone[(⊠)][iD]

Department of Computer Science, University of Pisa,
Largo B. Pontecorvo 3, 56127 Pisa, Italy
veronica.guerrini@di.unipi.it, giovanna.rosone@unipi.it

Abstract. The development of Next Generation Sequencing has had a major impact on the study of genetic sequences, and in particular, on the advancement of metagenomics, whose aim is to identify the microorganisms that are present in a sample collected directly from the environment. In this paper, we describe a new lightweight alignment-free and assembly-free framework for metagenomic classification that compares each unknown sequence in the sample to a collection of known genomes. We take advantage of the combinatorial properties of an extension of the Burrows-Wheeler transform, and we sequentially scan the required data structures, so that we can analyze unknown sequences of large collections using little internal memory. For the best of our knowledge, this is the first approach that is assembly- and alignment-free, and is not based on k-mers. We show that our experiments confirm the effectiveness of our approach and the high accuracy even in negative control samples. Indeed we only classify 1 short read on 5,726,358 random shuffle reads. Finally, the results are comparable with those achieved by read-mapping classifiers and by k-mer based classifiers.

Keywords: Metagenomics · Next-generation sequencing ·
Classification · Alignment-free · Assembly-free · eBWT · Lcp array

1 Introduction

The advent of "next-generation" DNA sequencing (NGS) technologies has meant that collections of hundreds of millions of DNA sequences are now commonplace in bioinformatics. One research field that has grown extraordinarily in recent years is metagenomics [26]. The problem of comparing sequences is of fundamental importance in this field: indeed, the metagenomic studies require computational tools that are able to analyze large datasets and to extract correct information about the community under investigation. There exist many metagenomics statistical/computational tools (among them [11,22,25,30]) and

VG is totally and GR is partially supported by the project MIUR-SIR CMACBioSeq ("Combinatorial methods for analysis and compression of biological sequences") grant n. RBSI146R5L.

© Springer Nature Switzerland AG 2019
I. Holmes et al. (Eds.): AlCoB 2019, LNBI 11488, pp. 112–124, 2019.
https://doi.org/10.1007/978-3-030-18174-1_8

recent surveys (for instance [13,21]) that offer a thorough benchmarking analysis by comparing the majority of the state-of-the-art tools.

We propose a new alignment-free and assembly-free method for comparing sequences (cf. [20,29]), which is combinatorial by nature and allows us to use little internal memory with respect to other approaches, such as those k-mer based. Our method is based on an extension of the Burrows-Wheeler Transform (shortly eBWT) to a collection of sequences. The eBWT has been used in several application contexts as the circular pattern matching (cf. [9]) and the alignment-free methods for comparing sequences (cf. [14,17,18,31]). Different distance measures have been defined and successfully applied to several biological datasets, as for instance mitochondrial DNA genomes [17,18], expressed sequence tags [23] and proteins [31]. Usually, when the eBWT is applied to a collection $\mathcal{S}$ of sequences, its output string ebwt($\mathcal{S}$) is enriched by another data structure: the document array da($\mathcal{S}$), i.e. a different color can be assigned to each element of $\mathcal{S}$ and each symbol of ebwt($\mathcal{S}$) is associated with a color in da($\mathcal{S}$). In other words, the array da($\mathcal{S}$) contains a sequence of colors that depends on how the suffixes of the sequences in $\mathcal{S}$ are mixed in the sorted list. In [17,18], the authors define a class of dissimilarity measures that, by using the eBWT, formalize the intuitive idea that the greater is the number of substrings shared by two sequences u and v, the smaller is the "distance" between u and v.

In this paper, inspired by the same intuitive idea, we define a new similarity measure that is based on an important property of the eBWT (the clustering effect [19,28] of the input symbols) together with the information on the length of the contexts that follow them. Then we describe how to apply this notion of similarity to perform metagenomic classifications.

Finally, in the last section, we describe the results of preliminary experiments on simulated metagenome collections: we obtain similar or better precision than CLARK-S [24] (a k-mer based approach), yet using a smaller memory footprint.

We show, moreover, that the sensitivity and precision obtained are similar to those achieved by Centrifuge [11], a read-mapping classifier. Nevertheless, our method gets better results than Centrifuge on the negative control datasets comprising random shuffled reads that do not belong to any known genome. Indeed, in metagenomic samples, a large number of reads are from "unknown" organisms whose genomes are not present in any reference database, and thus they cannot be given a taxonomic assignment. To mimic these reads, negative control datasets have been designed as to test the reliability of a method [13].

2 Preliminaries and Materials

Let S be a string (or sequence), n its length, and Σ its alphabet set, with $\sigma = |\Sigma|$. We denote the i-th symbol of S by $S[i]$. We denote by $\mathcal{S} = \{S_1, S_2, \ldots, S_m\}$ a collection of m strings. We assume that each string $S_i \in \mathcal{S}$ of length n_i is followed by a special symbol $S_i[n_i + 1] = \$_i$, which is lexicographically smaller than any other characters in $\mathcal{S}$, and do not appear in $\mathcal{S}$ elsewhere—for implementation purposes, we may simply use a unique end-marker $\$$ for all strings in $\mathcal{S}$. A *substring* of any $S \in \mathcal{S}$ is denoted as $S[i,j] = S[i] \cdots S[j]$, with $S[1,j]$ being called a *prefix* and $S[i, n+1]$ a *suffix* of S. A *range* is delimited by a square bracket if the correspondent endpoint is included.

The *Burrows-Wheeler Transform* (BWT) [4] is a well known reversible string transformation widely used that can be extended to a collection of strings. Such an extension, known as eBWT or multi-string BWT, is a reversible transformation that produces a string (denoted by $\mathsf{ebwt}(\mathcal{S})$) that is a permutation of the symbols of all strings in $\mathcal{S}$ [17] (see also [2,6,7,16]). The length of $\mathsf{ebwt}(\mathcal{S})$ is denoted by $N = \sum_{i=1}^{m} n_i + m$ and $\mathsf{ebwt}(\mathcal{S})[i] = x$, with $1 \leq i \leq N$, if x circularly precedes the i-th suffix $S_j[k, n_j + 1]$ (for some $1 \leq j \leq m$ and $1 \leq k \leq n_j+1$), according to the lexicographic sorting of the suffixes of all strings in $\mathcal{S}$. In this case we say that the suffix $S_j[k, n_j+1]$ is associated with the position i in $\mathsf{ebwt}(\mathcal{S})$ and with the color $j \in \{1, 2, \ldots, m\}$. The output string $\mathsf{ebwt}(\mathcal{S})$ is enhanced with the array $\mathsf{da}(\mathcal{S})$ of length N where $\mathsf{da}(\mathcal{S})[i] = j$, with $1 \leq j \leq m$ and $1 \leq i \leq N$, if $\mathsf{ebwt}(\mathcal{S})[i]$ is a symbol of the string $S_j \in \mathcal{S}$. See Fig. 1 for an example.

The *longest common prefix* (LCP) array of $\mathcal{S}$ is the array $\mathsf{lcp}(\mathcal{S})$ of length $N + 1$, such that $\mathsf{lcp}(\mathcal{S})[i]$, with $2 \leq i \leq N$, is the length of the longest common prefix between the suffixes associated with the positions i and $i - 1$ in $\mathsf{ebwt}(\mathcal{S})$, and $\mathsf{lcp}(\mathcal{S})[1] = \mathsf{lcp}(\mathcal{S})[N + 1] = 0$ by default.

The set $\mathcal{S}$ will be omitted if it is clear from the context. Moreover, for clarity of description, we denote by $\mathcal{L}(\mathcal{S})$ the sorted list of the suffixes in $\mathcal{S}$, although we do not need it for our computation.

We call *u-occurrence* any substring u that occurs in any sequence of $\mathcal{S}$.

Remark 1. Recall that $\mathsf{ebwt}(\mathcal{S})$ is implicitly associated with $\mathcal{L}(\mathcal{S})$ and all the suffixes in $\mathcal{S}$ starting with the same substring u, with $|u| = k$, must be consecutive entries in $\mathcal{L}(\mathcal{S})$ in the range $[h, j]$. Moreover, $\mathsf{lcp}[i] \geq k$ for $i = h + 1, \ldots, j$ and the symbols of $\mathcal{S}$ that are followed by u-occurrences coincide with $\mathsf{ebwt}[h, j]$.

Remark 2. Let ℓ be the total number of u-occurrences in $\mathcal{S}$, with $|u| = k$, there exist $k - 1$ substrings (*i.e.* all suffixes of u that are not equal to u) which appear *at least* ℓ times in $\mathcal{S}$.

Example 3 (running example). Let $\mathcal{S} = \{S_1 = \mathbf{ACGTCGCATTAA}, S_2 = \mathbf{CGTCACATNA}\}$. The substring CGT appears exactly once in both sequences. The two suffixes of S_1 and S_2 starting with CGT-occurrences occupy consecutive positions, precisely 14 and 15 in Fig. 1, and $\mathsf{lcp}[15] = 4$. Moreover, according to Remark 2 the number of GT-occurrences is 2 and the one of T-occurrences is 5.

3   Method

In this section, we introduce a new strategy to tackle the problem of metagenomic classification that is assembly- and alignment-free, not based on k-mer, and uses a little amount of memory.

We suppose that $\mathcal{S} = \{S_1, \ldots, S_m\}$ is a collection of biological sequences comprising r reads and g genomes, where $m = r + g$. More in details, $S_i \in \mathcal{S}$ is

index	$da(\mathcal{S})$	$lcp(\mathcal{S})$	$ebwt(\mathcal{S})$	$\mathcal{L}(\mathcal{S})$
1	1	0	A	$\$_1$
2	2	0	A	$\$_2$
3	1	0	A	$A\$_1$
4	2	1	N	$A\$_2$
5	1	1	T	$AA\$_1$
6	2	1	C	$ACATNA\$_2$
7	1	2	$\$_1$	$ACGTCGCATTAA\$_1$
8	2	1	C	$ATNA\$_2$
9	1	2	C	$ATTAA\$_1$
10	2	0	T	$CACATNA\$_2$
11	2	2	A	$CATNA\$_2$
12	1	3	G	$CATTAA\$_1$
13	1	1	T	$CGCATTAA\$_1$
14	2	2	$\$_2$	$CGTCACATNA\$_2$
15	1	4	A	$CGTCGCATTAA\$_1$
16	1	0	C	$GCATTAA\$_1$
17	2	1	C	$GTCACATNA\$_2$
18	1	3	C	$GTCGCATTAA\$_1$
19	2	0	T	$NA\$_2$
20	1	0	T	$TAA\$_1$
21	2	1	G	$TCACATNA\$_2$
22	1	2	G	$TCGCATTAA\$_1$
23	2	1	A	$TNA\$_2$
24	1	1	T	$TTAA\$_1$

Fig. 1. The required data structures for our running example, where $\mathcal{S}$ is the set $\{S_1 = ACGTCGCATTAA, S_2 = CGTCACATNA\}$

a read if $1 \leq i \leq r$ and $S_j \in \mathcal{S}$ is a genome if $r + 1 \leq j \leq m$. For simplicity, we denote by $\mathcal{R}$ the subset of reads and by $\mathcal{G}$ the subset of genomes. Assume that Σ is the biological alphabet of these sequences.

We introduce a method that classifies any read S_i in $\mathcal{R}$ by assigning it to a unique genome $S_j \in \mathcal{G}$ through reading in sequential way $ebwt(\mathcal{S}), da(\mathcal{S})$ and $lcp(\mathcal{S})$.

We define a notion of similarity between sequences that exploits the underlying properties of the eBWT: the clustering effect, i.e. the fact that this transformation tends to group together symbols that occur in similar contexts in the input string collection. Indeed, when applying eBWT to a collection, if a substring u occurs in one or more sequences, then the suffixes of the collection starting with u are likely to be close in the sorted list of suffixes. This implies that the greater the number of substrings shared by two sequences is, the more they are similar. Roughly speaking, we consider the symbols of $\mathcal{S}$ followed by a same substring (i.e. context) that are clustered together in $ebwt(\mathcal{S})$ and match one-to-one the symbols belonging to $\mathcal{R}$ to the symbols belonging to $\mathcal{G}$.

Our method works in three steps: (1) we detect and keep some blocks of $ebwt(\mathcal{S})$ in which the suffixes in $\mathcal{L}(\mathcal{S})$ share a common context of a minimum length α, and to which sequences both in $\mathcal{R}$ and $\mathcal{G}$ belong; (2) we analyze these interesting blocks in order to evaluate a degree of similarity between any read and any genome in $\mathcal{S}$; (3) we perform the read assignment: for every read in $\mathcal{R}$, either we retrieve the unique genome of belonging, or we report that it is not possible to identify it.

Build α-Clusters Collection—In step (1), inspired by Remark 1, we build a collection $\mathcal{C}_\alpha$ of blocks in $\mathsf{ebwt}(\mathcal{S})$, which are delimited by pairs of indices called α-clusters, that are associated with LCP-values exceeding a threshold value α.

Definition 4. *Let α be a positive integer, $\mathsf{lcp}[1, N+1]$ be the LCP-array and $\mathsf{da}[1, N]$ the document array associated with $\mathsf{ebwt}[1, N]$. An α-cluster of $\mathsf{ebwt}(\mathcal{S})$ of size $pE - pS + 1$ is any pair of indices (pS, pE) in $[1, N]$ such that*

- $\mathsf{lcp}[pS] < \alpha$, and $\mathsf{lcp}[pE + 1] < \alpha$,
- $\mathsf{lcp}[i] \geq \alpha$, for every $pS < i \leq pE$,
- *there exist two indices s, t, $pS \leq s, t \leq pE$, such that $\mathsf{da}[s] \leq r$ and $\mathsf{da}[t] > r$, where r is the total number of reads in $\mathcal{S}$.*

Example 5 (running example). For $\alpha = 2$, the set $\mathcal{C}_2$ of 2-clusters of the $\mathsf{ebwt}(\mathcal{S})$ in Fig. 1 is $\mathcal{C}_2 = \{(6, 7), (8, 9), (10, 12), (13, 15), (17, 18), (21, 22)\}$.

In other words, we discard the blocks of $\mathsf{ebwt}(\mathcal{S})$ whose associated suffixes do not share a prefix of length at least α. This step requires a sequential scan of $\mathsf{lcp}(\mathcal{S})$ and $\mathsf{da}(\mathcal{S})$ allowing us to use only a small amount of memory to detect α-clusters.

Remark 6. It is easy to see that we are computing the similarity between a read $S_j \in \mathcal{R}$ and a genome $S_k \in \mathcal{G}$ by analyzing the entire set of sequences $\mathcal{S}$, not only the two sequences S_j and S_k. Indeed, let (pS, pE) be an α-cluster of $\mathsf{ebwt}(\mathcal{S})$ that contains at least a symbol of S_j and at least a symbol of S_k. Other symbols that belong to sequences in $\mathcal{S}$ apart from S_j and S_k may also appear in $\mathsf{ebwt}[pS, pE]$. Nevertheless, we can implicitly get a new cluster (pS', pE') by deleting from the $\mathsf{ebwt}(\mathcal{S})$ all symbols not belonging to S_j and S_k, and for the properties of the LCP array, it is easy to verify that (pS', pE') forms an α-cluster.

Build Similarity Arrays—During the second step, we refine each α-cluster of the $\mathsf{ebwt}(\mathcal{S})$ by splitting it according to its symbols, and then we measure the degree of similarity between the sequences in $\mathcal{R}$ and the genomes in $\mathcal{G}$.

We split the alphabet Σ of $\mathcal{S}$ in two subsets. We include the DNA bases in $\Sigma' = \{A, C, G, T\}$ and the end-marker symbols, the (rare) occurrences of N and other degenerate base symbols (see IUPAC nomenclature) in $\Sigma'' = \Sigma \setminus \Sigma' \cup \{\$\}$.

Definition 7. *Let a be any symbol in Σ'. The a-refinement of an α-cluster (pS, pE) of $\mathsf{ebwt}(\mathcal{S})$ is the set of indices $\{j_1, \ldots, j_q\}$ in the range $[pS, pE]$, such that $\mathsf{ebwt}[j_\ell] = a$, for any $1 \leq \ell \leq q$.*

Example 8 (running example). The C-refinement of the cluster $(6, 7) \in \mathcal{C}_2$ is the singleton $\{6\}$, while the a-refinement, for any $a \in \{A, G, T\}$, is the empty set, since neither A nor G nor T appear in $\mathsf{ebwt}[8, 9]$. On the other hand, the C-refinement of the cluster $(8, 9) \in \mathcal{C}_2$ is the set $\{8, 9\}$.

Now, we define a similarity between two sequences $S_j, S_k \in \mathcal{S}$ by using the notion of α-cluster and a-refinement.

Definition 9. *Let C_α be the set of all the α-clusters associated with* ebwt(S). *We define the α-similarity between two sequences $S_j \in \mathcal{R}$ and $S_k \in \mathcal{G}$ as the quantity* $\mathfrak{S}_\alpha(S_j, S_k) = \sum_{x \in C_\alpha} Q_{j,k}(x)$, *where*

$$Q_{j,k}(x) = \sum_{a \in \Sigma'} \min \left(n_{(j,x,a)}, n_{(k,x,a)} \right), \tag{1}$$

with $n_{(j,x,a)}$ (resp. $n_{(k,x,a)}$) being the number of symbols belonging to S_j (resp. S_k) in the a-refinement of the α-cluster x.

Intuitively, during the computation of our measure, we count the symbols of each read that we can associate with the same symbols of each genome in the α-cluster, or vice versa. In particular, if the symbol belongs to Σ', we associate the nucleotide of a read in $\mathcal{R}$ with the exact nucleotide of a genome in $\mathcal{G}$. Whereas if the symbol belongs to Σ'', we consider it as placeholder, in the sense that we associate it with any nucleotide of the sequence of the other collection in order to maximize the quantity $Q_{i,j}(x)$ in Eq. (1) (see Example 10 below).

More precisely, let $m_{(j,x)}$ (resp. $m_{(k,x)}$) be the number of Σ''-symbols belonging to S_j (resp. S_k) and appearing in an α-cluster x. For any $a \in \Sigma'$, if $|n_{(j,x,a)} - n_{(k,x,a)}| > 0$ (i.e. the minimum between the number of a-symbols of S_j appearing in x and the number of a-symbols of S_k appearing in x can be increased), then we convert some placeholders to a-symbols and decrease the quantities $m_{(j,x)}$ and $m_{(k,x)}$ accordingly. Note that the symbol $a \in \Sigma'$ to which we convert any placeholder symbol appearing in x is unique.

Furthermore, if n_j (resp. n_k) is the length of S_j (resp. S_k), then the quantity $\mathfrak{S}_\alpha(S_j, S_k)$ ranges between 0 and $\min(n_j, n_k) + 1 - \alpha$. We can normalize dividing by $\min(n_j, n_k) + 1 - \alpha$, as to obtain a similarity value within the range $[0, 1]$.

Example 10 (running example). The 2-similarity between S_1 and S_2 is given by $\mathfrak{S}_2(S_1, S_2) = 1 + 1 + 0 + 1 + 1 + 1 = 5$, by setting $\$_1 = C$ and $\$_2 = T$, and by normalizing $\mathfrak{S}_2(S_1, S_2)/9 = 0,56$. On the other hand, for $\alpha = 3$, the normalized similarity $\mathfrak{S}_3(S_1, S_2)$ is equal to $0,25$ if and only if $\$_2 = A$.

Concerning the metagenome analysis and the set S, we build a set of similarity arrays $\{Sim_1, \ldots, Sim_g\}$. More precisely, for each genome $S_k \in \mathcal{G}$, we define the array Sim_{k-r} of length r, whose entry $Sim_{k-r}[j]$ stores the normalized similarity value $\mathfrak{S}_\alpha(S_j, S_k)$, for every $S_j \in \mathcal{R}$.

In order to use only a sequential scan of ebwt(S) and da(S), we analyze the α-clusters in C_α one by one through $|C_\alpha|$ iterations. At each iteration, we consider an α-cluster x and we evaluate the quantities $Q_{j,k}(x)$ of Eq. (1), for every index $j \leq r$ and $k > r$ appearing in x, maximizing them by means of placeholder symbols (if there are any). Then, we update each corresponding entry $Sim_{k-r}[j]$ by adding the quantity $Q_{j,k}(x)$.

Finally, once all the α-clusters in C_α have been examined, we normalize each entry of $\{Sim_1, \ldots, Sim_g\}$ completing the construction of the similarity arrays.

Classification—The last step consists in assigning a unique provenance to any read S_j ($j \leq r$) with respect to the normalized values $\mathfrak{S}_\alpha(S_j, S_k)$, $r < k \leq r + g$.

For every $j \leq r$, we compute the set $\mathcal{I}$ of indices q such that the normalized similarity $\mathfrak{S}_\alpha(S_j, S_q)$ is close to the maximum normalized similarity $M = \max_i \mathfrak{S}_\alpha(S_j, S_i)$, i.e.

$$\mathfrak{S}_\alpha(S_j, S_q) \sim M, \text{ for all } q \in \mathcal{I}. \tag{2}$$

Moreover, in order to control the assessment score, we set a threshold value β ($0 \leq \beta < 1$) that the maximum value M of Eq. (2) has to exceed in order to classify the read S_j with respect to $\mathcal{I}$.

We assign the read S_j (or its reverse complement) to S_q if $q \in \mathcal{I}$, $|\mathcal{I}| = 1$ and $\mathfrak{S}_\alpha(S_j, S_q) > \beta$. Whereas, the read S_j is said to be *not classified* if $\max_i \mathfrak{S}_\alpha(S_j, S_i) \leq \beta$. Finally, the read classification of S_j is said to be *ambiguous* if $\mathfrak{S}_\alpha(S_j, S_q) > \beta$ and $|\mathcal{I}| > 1$. In the last case, if our strategy is used for the analysis of a paired-end collection, we use the sum of the assignment scores of the individual mates and assign the read to the genome that obtains the maximum score. Note that if more than one genome obtains the maximum score, we could classify the read at higher taxonomic ranks.

4   Results

In this section we evaluate our alignment-free strategy against other tools. We choose two tools: the first is alignment-free and is based on the use of k-mers, and the second is based on a read-mapping strategy. To assess the performance of our sequence analysis method, we have implemented a prototype C++ tool, named LightMetaEbwt[1].

A recent evaluation of the state-of-the-art tools for metagenome classification [13] presents the most widely used tools tested on complex and realistic datasets which have been designed ad hoc for this analysis[2]. According to this benchmarking analysis, kraken [30] and CLARK [25] result to be top-performing tools in terms of both similarity to the correct answer and the fraction of reads classified [13]. Note that both tools are k-mer based. However, for our evaluation, we selected the new version of CLARK, called CLARK-S [24], that uses spaced k-mers rather than simple k-mers, and achieves higher sensitivity than both CLARK and kraken, while maintaining high precision. Nevertheless, the tool CLARK-S, as well as CLARK and kraken, is extremely memory-consuming, and the results obtained by running its lightweight version CLARK-l are indicated to be a "draft, or first order approximation" of those obtained by running CLARK or CLARK-S.

We also compare our results with a recent metagenomics classifier, named Centrifuge [11]. It adapts the data structures of read-mapping algorithms based on the BWT and the FM-index [8], such as Bowtie [12], which provide very fast alignment with a relatively small memory footprint. We observe that LightMetaEbwt, unlike Centrifuge, processes all reads at the same time.

---

[1] https://github.com/veronicaguerrini/LightMetaEbwt.
[2] http://www.gardner-binflab.org/our_research/.

In order to guarantee a fair evaluation, we use the custom reference database for the three tools. Notice that a like-for-like comparison on the time-consuming between LIGHTMETAEBWT, CLARK-S and Cenfrifuge is not possible, since CLARK-S and Centrifuge are multi-thread and our tool is currently able to use one core only. In order to run CLARK-S, we use a machine with 128 GB of RAM. All tests were done on a DELL PowerEdge R630 machine, 24-core machine with Intel(R) Xeon(R) CPU E5-2620 v3 at 2.40 GHz, with 128 GB of shared memory, used in not exclusive mode. The system is Ubuntu 14.04.2 LTS.

Dataset Description. The reference database $\mathcal{G}$ we use for our experiments comprises 930 genomes from 686 species belonging to 17 phyla as indicated in [13].

We perform validation of our approach by using two sets of metagenomes among those provided by Lindgreen et al. [13]: the two datasets of paired end reads *setA2* and *setB2* reproduce the size, complexity and characteristics of real metagenomic samples containing around 20 millions of sequences of length 100 belonging to $\mathcal{R}$. Some phyla are included in equal proportions, whereas some others vary more substantially between the two sets.

Moreover, as to test the reliability of the tools, each dataset includes a subset of simulated negative control sequences to mimic sequences from "unknown" organisms (i.e. their genomes are not present in the reference database) that are likely to appear in metagenome samples – see [13] for further details. Each of these negative control datasets, called *setA2_Ran* and *setB2_Ran* in our experiments, includes around 5 million of random shuffled reads.

We precise that the original datasets, downloadable from [13], are not exactly the datasets *setA2* and *setB2* we use for our evaluations[3]. In fact, we first removed a group of reads associated with the phylum of Eukaryotes whose species provenance was not specified in [13]. Second, since we use a custom database and CLARK-S downloads up-to-date taxonomy data (such as taxonomy id, or accession numbers) from the NCBI website ignoring expiring entries, we preferred not to include in sets *setA2* and *setB2* a group of reads associated with 3 genomes whose entries in the NCBI database have been indicated as obsolete.

Preprocessing Step. This task for our tool can be achieved using, for example, BCR [5], Egsa [16], gsacak [15], GAP [7] or eGAP [6]. As the set $\mathcal{G}$ of genomes is the same for each experiment, we can build the data structures of $\mathcal{G}$ only once, by using GAP[4]. Then we can use BCR[5] (it is a tool for very large collection of short reads) for building the data structures for $\mathcal{R}$ and use eGAP[6] for merging them obtaining the data structures for the entire collection $\mathcal{S}$. On the other hand, exploiting the mathematical properties of the permutation associated with the eBWT and LCP array, by using BCR [2, Remark 3.6], [5], we can update the data structures of $\mathcal{G}$ (without constructing the BWT from scratch) in order to obtain the data structures for $\mathcal{S}$. To find the best method for building our data structures is a non-trivial problem and it is not in the aim of this paper.

---

[3] https://github.com/veronicaguerrini/LightMetaEbwt/tree/master/Datasets.
[4] http://people.unipmn.it/manzini/gap.
[5] https://github.com/giovannarosone/BCR_LCP_GSA.
[6] https://github.com/felipelouza/egap.

Table 1. Results of metagenome analysis at species level of *setA2* and *setB2* for positive control, and *setA2_Ran* and *setB2_Ran* for negative control. Best scores are in bold

	REAL	CLARK-S highconfidence	LightMetaEbwt α 16 β 0.25	LightMetaEbwt α 16 β 0.35	Centrifuge min-hitlen 16	Centrifuge min-hitlen 22
setA2						
TP	21,461,160	19,789,944	19,908,394	19,815,751	20,062,940	19,897,787
FP	0	187,386	37,232	34,408	485,353	68,722
FN	0	1,483,830	1,515,534	1,611,001	912,867	1,494,651
SEN (%)	100.000	93.025	92.926	92.481	**95.648**	93.013
PREC (%)	100.000	99.062	99.813	**99.827**	97.638	99.656
F1 (%)	100.000	95.949	**96.247**	96.014	**96.633**	96.220
setB2						
TP	20,249,373	18,644,316	18,922,266	18,819,348	18,913,373	18,766,021
FP	0	167,709	73,208	68,154	450,209	58,766
FN	0	1,437,348	1,253,899	1,361,871	885,791	1,424,586
SEN (%)	100.000	92.842	93.785	93.251	**95.526**	92.944
PREC (%)	100.000	99.109	99.615	**99.639**	97.675	99.688
F1 (%)	100.000	95.873	**96.612**	96.340	96.589	96.198
setA2_Ran						
TN	5,726,358	5,726,336	5,726,294	5,726,357	150,971	5,712,085
FP	0	22	64	1	5,575,387	14,273
SPEC (%)	100.00	99.99	99.99	**100.00**	2.64	99.75
setB2_Ran						
TN	5,406,659	5,406,642	5,406,601	5,406,658	141,994	5,393,260
FP	0	17	58	1	5,264,665	13,399
SPEC (%)	100.00	99.99	99.99	**100.00**	2.63	99.75

Validation Step. As the provenance of simulated reads is known, we can set TP as the number of reads correctly classified (i.e. assigned to their right provenance), FP as the number of reads erroneously classified, and FN as the number of reads unassigned, from which we can calculate the quality metrics: sensitivity $SEN = \frac{TP}{TP+FN}$, precision $PREC = \frac{TP}{TP+FP}$, and F1 score $F_1 = \frac{2TP}{2TP+FP+FN}$. In these experiments, we do not handle the reads classified as ambiguous or assigned to taxonomic level higher than species (i.e. more species could be assigned to them), and we count them among unclassified reads in FN. For simulated negative control sequences that do not exist in any known species, we can set TN as the number of random reads that are not mapped to any genome and FP as the number of random reads that are erroneously mapped to some genome, and calculate the specificity $SPEC = \frac{TN}{TN+FP}$.

Experiments. Our tool is able to classify the reads to several taxonomic levels such as genomes, species or phylum. For the experiments reported in Table 1, we choose a deep taxonomic level, i.e. we classify each read to the species level.

CLARK-S runs with default values and the results are filtered by using the recommended option `--highconfidence` (e.g., assignment with confidence score <0.75 and gamma score <0.03 are discarded).

For LightMetaEbwt, we set the minimum length of the common context $\alpha = 16$, since the length of each paired end read is 100, and provide results for minimum similarity scores $\beta = 0.25$ and $\beta = 0.35$. Our similarity score ranges between 0 and 1: clearly the greater the value is, the higher the read similarity is. Thus, for $\beta = 0.25$ the sensitivity increases and the precision slightly decreases.

Centrifuge begins with a short exact match (16-bp minimum) and extends the match as far as possible. Based on the exact matches found in the read and its reverse complement, Centrifuge classifies each read using only those mappings with at least one match of k bases. This parameter k (named `--min-hitlen`) is comparable with α used in our tool. Hence, we perform a first experiment where we set it to 16 and a second experiment to 22 (default value). For both *setA2* and *setB2*, the highest sensitivity achieved is given by choosing `--min-hitlen`=16. Nevertheless, for `--min-hitlen`=16 such a higher sensitivity alters the correct metagenomic classification as the percentage of random shuffled reads classified (i.e. the specificity) dramatically decreases up to 2.63%. The higher sensitivity for `--min-hitlen`=16 increases the F1 score, which is the harmonic mean of precision and sensitivity. In fact, for *setA2* the best F1 score is obtained by Centrifuge for `--min-hitlen`=16 followed by LIGHTMETAEBWT with $\beta = 0.25$. Further experiments with $\beta = 0.1$ show that the F1 score obtained by LIGHT-METAEBWT increases to 97.1% at the cost of a slightly low specificity (98.4%).

Without considering the pre-processing steps, we observe that the RAM usage of our tool (by using a semi-external memory approach) is about 17–18 GB for *setA2* and *setB2* and about 9–10 GB for *setA2_Ran* and *setB2_Ran*. CLARK-S uses about 120 GB for any dataset, whereas Centrifuge uses less than 2 GB. Moreover, we observe that our method scans sequentially the required data structures, so that we could analyze unknown sequences of large collections in external memory by reducing the internal memory usage to a minimum. We have also observed that our tool is slower than the other two tools, but a more engineered implementation of our algorithm would improve our performance in terms of time and space, that we leave as further work.

Overall accuracy for the three tools was very similar, but the highest precision (keeping high sensitivity) values are obtained by our tool even in the random shuffled samples (*setA2_Ran* and *setB2_Ran*).

5 Conclusions and Discussion

In this paper, we present a versatile, alignment-free, lightweight method that by sequentially scanning some data structures eBWT, LCP and DA array allows us to identify the genome to which each read belongs. We focused the attention on species level classifications, but LIGHTMETAEBWT can also work at higher taxonomic levels such as genus, family, class or phylum. Preliminary experiments show that the relative phylum abundance estimated meets the real dataset composition with very high precision. For instance, we obtain only $31,666$ ambiguous reads and $868,456$ not classified reads and we correctly classify $19,349,193$ in *setB2*.

Furthermore, we have considered the sequences classified as ambiguous as those not classified, but we leave a more in-depth analysis of the ambiguous reads for a further work, for instance using our tool with stronger parameters.

The idea of building the clusters of the eBWT with/without LCP is not new (see [17,18]). However, we want to specify that our notion differs from the notion of LCP-interval in the literature [1]: indeed, a LCP-interval is a particular α-cluster (pS, pE) in which at least an index i, $pS < i \leq pE$, is equal to α. It is well worth mentioning also the difference with the strategies used in [10,27], where the partitions of ebwt($\mathcal{S}$) determined by LCP-values are filtered according to their size. Here, we do not impose any constraint on the α-cluster size. Moreover, to the best of our knowledge, it is the first time that the notion of cluster is used on metagenomic classification problems.

Furthermore, it is interesting to note that the data structures used by our strategy are intrinsically dynamic: the collection $\mathcal{S}$ can be modified by inserting or removing sequences [2, Remark 3.6], [5] exploiting the mathematical properties of the permutation associated with the eBWT and LCP array (for instance by using BCR), allowing us to modify α-clusters accordingly. On the other hand, one can build and store the data structures for the genome database and then, for each new experiment, build the data structures for the read collection. To merge them and obtain ebwt($\mathcal{S}$), da($\mathcal{S}$) and lcp($\mathcal{S}$), one could use eGAP.

Finally, note that LIGHTMETAEBWT allows a certain degree of parallelization. The analysis of the clusters is independent of each other and is thus easily parallelizable. This allows us to use multiple processors on multi-core servers that are commonplace nowadays while keeping the computational requirements low. Moreover, we note that in the recent literature there are several papers with the aim of introducing new lightweight and parallel computational strategies for building the data structures we use in our tool, for instance see [3].

In conclusion, we believe that our tool can be useful in a variety of applications in metagenomics and genomics.

References

1. Abouelhoda, M.I., Kurtz, S., Ohlebusch, E.: Replacing suffix trees with enhanced suffix arrays. J. Discrete Algorithms **2**(1), 53–86 (2004)
2. Bauer, M., Cox, A., Rosone, G.: Lightweight algorithms for constructing and inverting the BWT of string collections. Theoret. Comput. Sci. **483**, 134–148 (2013)
3. Bonizzoni, P., Della Vedova, G., Nicosia, S., Pirola, Y., Previtali, M., Rizzi, R.: Divide and conquer computation of the multi-string BWT and LCP array. In: Manea, F., Miller, R.G., Nowotka, D. (eds.) CiE 2018. LNCS, vol. 10936, pp. 107–117. Springer, Cham (2018). https://doi.org/10.1007/978-3-319-94418-0_11
4. Burrows, M., Wheeler, D.: A block sorting data compression algorithm. Technical report, DIGITAL System Research Center (1994)
5. Cox, A., Garofalo, F., Rosone, G., Sciortino, M.: Lightweight LCP construction for very large collections of strings. J. Discrete Algorithms **37**, 17–33 (2016)
6. Egidi, L., Louza, F.A., Manzini, G., Telles, G.P.: External memory BWT and LCP computation for sequence collections with applications. In: WABI 2018. LIPIcs, vol. 113, pp. 10:1–10:14 (2018)
7. Egidi, L., Manzini, G.: Lightweight BWT and LCP merging via the gap algorithm. In: Fici, G., Sciortino, M., Venturini, R. (eds.) SPIRE 2017. LNCS, vol. 10508, pp. 176–190. Springer, Cham (2017). https://doi.org/10.1007/978-3-319-67428-5_15

8. Ferragina, P., Manzini, G.: Opportunistic data structures with applications. In: FOCS, pp. 390–398 (2000)
9. Hon, W.-K., Ku, T.-H., Lu, C.-H., Shah, R., Thankachan, S.V.: Efficient algorithm for circular Burrows-Wheeler transform. In: Kärkkäinen, J., Stoye, J. (eds.) CPM 2012. LNCS, vol. 7354, pp. 257–268. Springer, Heidelberg (2012). https://doi.org/10.1007/978-3-642-31265-6_21
10. Janin, L., Rosone, G., Cox, A.J.: Adaptive reference-free compression of sequence quality scores. Bioinformatics 30(1), 24–30 (2014)
11. Kim, D., Song, L., Breitwieser, F.P., Salzberg, S.L.: Centrifuge: rapid and sensitive classification of metagenomic sequences. Genome Res. 26(12), 1721–1729 (2016)
12. Langmead, B., Trapnell, C., Pop, M., Salzberg, S.L.: Ultrafast and memory-efficient alignment of short DNA sequences to the human genome. Genome Biol. 10(3), R25 (2009)
13. Lindgreen, S., Adair, K.L., Gardner, P.P.: An evaluation of the accuracy and speed of metagenome analysis tools. Sci. Rep. 6, Article No. 19233 (2016)
14. Louza, F.A., Telles, G.P., Gog, S., Zhao, L.: Computing Burrows-Wheeler similarity distributions for string collections. In: Gagie, T., Moffat, A., Navarro, G., Cuadros-Vargas, E. (eds.) SPIRE 2018. LNCS, vol. 11147, pp. 285–296. Springer, Cham (2018). https://doi.org/10.1007/978-3-030-00479-8_23
15. Louza, F., Gog, S., Telles, G.: Inducing enhanced suffix arrays for string collections. Theor. Comput. Sci. 678, 22–39 (2017)
16. Louza, F., Telles, G., Hoffmann, S., Ciferri, C.: Generalized enhanced suffix array construction in external memory. Algorithms Mol. Biol. 12(1), 26 (2017)
17. Mantaci, S., Restivo, A., Rosone, G., Sciortino, M.: An extension of the Burrows-Wheeler transform. Theoret. Comput. Sci. 387(3), 298–312 (2007)
18. Mantaci, S., Restivo, A., Rosone, G., Sciortino, M.: A new combinatorial approach to sequence comparison. Theory Comput. Syst. 42(3), 411–429 (2008)
19. Mantaci, S., Restivo, A., Rosone, G., Sciortino, M., Versari, L.: Measuring the clustering effect of BWT via RLE. Theoret. Comput. Sci. 698, 79–87 (2017)
20. Mantaci, S., Restivo, A., Sciortino, M.: Distance measures for biological sequences: some recent approaches. Int. J. Approx. Reason. 47(1), 109–124 (2008)
21. McIntyre, A.B.R., et al.: Comprehensive benchmarking and ensemble approaches for metagenomic classifiers. Genome Biol. 18(1), 182 (2017)
22. Menzel, P., Ng, K.L., Krogh, A.: Fast and sensitive taxonomic classification for metagenomics with Kaiju. Nature Commun. 7, 11257 (2016)
23. Ng, K.H., Ho, C.K., Phon-Amnuaisuk, S.: A hybrid distance measure for clustering expressed sequence tags originating from the same gene family. PLoS One 7(10), e47216 (2012)
24. Ounit, R., Lonardi, S.: Higher classification sensitivity of short metagenomic reads with CLARK-S. Bioinformatics 32(24), 3823–3825 (2016)
25. Ounit, R., Wanamaker, S., Close, T.J., Lonardi, S.: CLARK: fast and accurate classification of metagenomic and genomic sequences using discriminative k-mers. BMC Genomics 16(1), 236 (2015)
26. Pedersen, M., et al.: Ancient and modern environmental DNA. Philos. Trans. R. Soc. Lond. B Biol. Sci. 370(1660), 20130383 (2015)
27. Prezza, N., Pisanti, N., Sciortino, M., Rosone, G.: Detecting mutations by eBWT. In: WABI 2018. LIPIcs, vol. 113, pp. 3:1–3:15 (2018)
28. Restivo, A., Rosone, G.: Balancing and clustering of words in the Burrows-Wheeler transform. Theoret. Comput. Sci. 412(27), 3019–3032 (2011)
29. Vinga, S., Almeida, J.: Alignment-free sequence comparison-a review. Bioinformatics 19(4), 513–523 (2003)

30. Wood, D.E., Salzberg, S.L.: Kraken: ultrafast metagenomic sequence classification using exact alignments. Genome Biol. **15**(3), R46 (2014)
31. Yang, L., Zhang, X., Wang, T.: The Burrows-Wheeler similarity distribution between biological sequences based on Burrows-Wheeler transform. J. Theor. Biol. **262**(4), 742–749 (2010)

MULKSG: *MUL*tiple *K* *S*imultaneous *G*raph Assembly

Christopher Wright[1]([✉]) [ID], Sriram Krishnamoorty[2] [ID], and Milind Kulkarni[1] [ID]

[1] Purdue University, West Lafayette, IN 47905, USA
{christopherwright,milind}@purdue.edu
[2] Pacific Northwest National Laboratory, Richland, WA 99352, USA
sriram@pnnl.gov

Abstract. This work shows how to parallelize multi K de Bruijn graph genome assembly simultaneously, removing the bottleneck of iterative multi K assembly. The expected execution time on a single node with 40 cores is variable, with the average execution time for the entire pipeline over 16 datasets tested being 1613 s for SPAdes vs. 1581 s for MULKSG, with the MULKSG graph creation and traversal averaging 15% faster than SPAdes. We implement a multi-node implementation for the graph creation and traversal portions of the assembly, showing the speedups in Fig. 4. We show that when implemented correctly with correction phases performed per graph in parallel, the expected outcome is very close to the original method, in some cases having less errors while keeping the same NGA50 and genome coverage %. We show this works in practice, implementing with the popular genome assembler SPAdes. Further, this algorithmic change gets rid of the single node sequential bottleneck on multi K genome assembly, allowing for the use of parallel error correction, graph building, graph correction, and graph traversal. We implement a parallel version of the assembly and show the statistics are the same as when run on a single node. The code is open source and can be found at https://github.com/cwright7101/mulksg.

Keywords: Genome assembly · Iterative assembly ·
Parallel de Bruijn graph · Multi K assembly

1 Introduction

Genome assembly is a widely studied field in bioinformatics, with algorithms and techniques evolving and changing as sequencing technology changes. Sequencing will produce raw strings of characters (base pairs or bp) called sequences or reads. These sequences range from hundreds of base pairs per sequence, to thousands or millions of bp per sequence. Genome assembly attempts to put these sequences in the right order, overlapping them to generate the "real" genome sequence. There is still no optimal algorithm or tool to assemble every genome.

Even small changes in assembly and error correction algorithms can have an enormous impact on the final output data. Making the data more accurate,

© Springer Nature Switzerland AG 2019
I. Holmes et al. (Eds.): AlCoB 2019, LNBI 11488, pp. 125–136, 2019.
https://doi.org/10.1007/978-3-030-18174-1_9

even if only correcting a few bases in every sequence in a database would be a huge win for current and future research. This leads to the motivation for the paper—where errors are and how to get rid of them while maintaining execution time. The most popular techniques for genome assembly include Overlap Layout Consensus (OLC) [26,29,34] and de Bruijn Graph(DBG) [25,30,31,33,35–37]. Both of these techniques will build a graph from the sequences, but their efficiency and accuracy are used in different situations. Both of these techniques are discussed in Sect. 2.

During first generation sequencing, OLC was the major assembly technique. In the mid 2000s, Next Generation Sequencing (NGS), commonly referred to as second generation sequencing, emerged. This technology produces sequences of 100–300 bp long, with millions or billions of reads for the entire genome. Because of space and time complexity of OLC, the DBG algorithm catapulted to popularity.

The DBG approach is faster and uses much less memory than the OLC approach. The problem is it is not as accurate. This motivated iteratively building different sized DBG, using smaller DBGs to build more accurate larger DBGs by combining the results [35]. Unfortunately, as more sizes of DBG are used during assembly, there is a linear increase in the amount of time taken for assembly. This limited the number of DBGs used, and hence the ultimate accuracy, irrespective of the amount of resources available.

The main insight of this work is that we show how to get rid of the sequential bottleneck that is inherent in iterative DBG assembly. Traditional DBGs take the input reads and split them into overlapping Kmers, or K-length keys/sub-strings. These keys are inserted into a graph as nodes, where Kmers are paired with their reverse complement in a single node. Each node inserted also keeps track of which nodes it is paired with in the paired end reads. Edges connect nodes when the two node keys are sequential in an input read, meaning $k - 1$ characters are the same in the two keys with the $k + 1$ characters spanning the two reads existing in the input read. For iterative assembly a first graph, with small k_1, is built, traversed to generate higher-quality k_2-mers (with $k_2 > k_1$) than would have existed in the original input. Then another graph is built from these derived k_2-mers, and traversed to produce new, larger Kmers. This process continues for increasing values of k until the desired final DBG is built. Our algorithm removes this iteration and instead performs a 2 phase building/traversal. We build $k - 1$ graphs in parallel, do the error correction and graph traversal to generate contigs. These contigs are passed to a final stage of building and traversal.

This is useful for accurate and fast assembly as the amount of genomic data generated is increasing at an exponential rate. With some sequencing centers generating 10–30 TB of data each day, the algorithms and stages of bioinformatics need to be parallel, efficient, and accurate. For a large portion of research areas, assembly gives the final sequences that are used downstream, and hence is the focus of this work. The paper first gives some background on genome assembly and techniques followed by a description of the implementation. Evaluation, related work and future work are presented before the conclusion.

2 Background

Current assembly tools can be categorized in a myriad of different ways. The tools could be categorized by the input data length (short read, long read, hybrid), the number of Kmer (K) values used during assembly (a single K value, or multiple K values), the algorithm type (overlap layout consensus, de Bruijn graph), or they can be categorize by the level of parallelism (single-thread, multi-thread, multi-node).

2.1 Overlap Layout Consensus

OLC will build an overlap graph, comparing all the different reads to each other and "overlap" the reads – pairing similar parts of different reads together. Different techniques of overlap might use suffix trees, dynamic programming, FM index, or inverted sub-string indexes. Common techniques to cut down on the time for this step include pruning the comparison space by only looking for overlaps where the reads have one or more sub-strings of some k length in common.

After building the overlap graph, OLC will then "layout" the graph, bundling the overlaps together and send these bundles to the "consensus" or error correction stage of assembly. The error correction or consensus then will pick the most likely sequence for each bundle, generating the output contigs. To pick these "most likely" sequences, usually a count or coverage statistic will be used.

While OLC is an accurate and matured technique, other approaches were searched for because of the long execution time and the large amount of memory needed during the overlap stage of assembly.

2.2 de Bruijn Graph Assembly

In de Bruijn Graph Assembly, the input reads are usually corrected (base correction and/or trimming). Next the graph is built using Kmers as mentioned in Sect. 1. In the DBG, a given node has a constant finite limit to the number of nodes that it has edges to. This graph then goes through simple error correction, usually based on frequency counts of the keys for each node. Next, paths in the graph that are unambiguous (in and out degree of the node are both equal to 1), are merged into a single node. Next, more error correction is done, usually tip clipping and bulge removal/co-removal. Finally, the graph is traversed to create contiguous sequences, or contigs.

During traversal, paired-end information is used to help determine which path to traverse in the graph. Some algorithms will stop at some ambiguities, where others are more aggressive and join the sequences together. Once the contigs are generated, a scaffolding phase is employed, where libraries of mate-pair reads are used to put the contigs into the right order and orientation. Some algorithms then do a post processing stage where polishing and error correction is done again.

In current multiple K value algorithms, the algorithm is iterative, starting at the lowest value of K, and increasing to the highest K. The process begins in the same manner as the single K value case, but after generating contigs for the lowest value K, instead of scaffolding, the algorithm takes the contigs and uses those as inputs to build the next biggest K value graph along with the input reads. The algorithm continues iteratively until it reaches the highest value K, then the algorithm continues with scaffolding as in the single K value case.

Popular multi K iterative algorithms include SPAdes [25], SOAPdenovo2 [31], IDBA-UD [35], MEGAHIT [30], and ScalaDBG [33]. SPAdes, IDBA-UD, SOAPdenovo2, and MEGAHIT are single node solutions, parallelized onto multiple cores on a single machine, with MEGAHIT available to use a GPU (CUDA) if available on the machine. ScalaDBG is either a single node or multi-node solution. The accuracy of ScalaDBG is similar to IDBA-UD as that is the backbone of the program. The ScalaDBG solution still has a serial bottleneck, as there are at least Log(K) levels of patching required in their algorithm, with the accuracy being hit or miss depending on the genome assembled.

With de Bruijn graph algorithms being prominent for second generation sequencing because of the accuracy, speed and memory usage, this work focuses on algorithms that use de Bruijn graph techniques, and further categorizes by algorithms that use a single K value versus multiple K values. As seen in SPAdes [25] and IDBA-UD [35], using multiple K values usually gives better accuracy than other techniques. This is intuitive, as one would expect that as more information is used, there would be a better assembly.

2.3 Existing Solutions

In de Bruijn graph algorithms, to improve on accuracy means improving the correction phase and how this data propagates. For IDBA-UD, MEGAHIT, and ScalaDBG, they use correction after the graph creation, assuming that read correction has already been performed. For SOAPdenovo2 and SPAdes, there is a pre-graph correction of the reads, even if the reads have been pre-processed, followed by graph creation and more error correction.

In preliminary discovery, using the newest available version of QUAST (v5.0.2) [28] to compare the multi-K assemblers on the GAGE-B [32] datasets, SPAdes was almost always the best in terms of genome fraction, N50, NG50, and NGA50, and all tools were relatively close in terms of miss-assemblies. SPAdes is also more well documented, has more recent development work, and seems to be the most widely used tool of the above mentioned. MEGAHIT focuses on metagenomes, whereas the others work on regular or metagenomes, hence this work focused on how SPAdes worked and reduced errors.

In SPAdes, the iterative procedure of building the de Bruijn/assembly graphs, is to start at the lowest K value, generate contigs, then use those contigs along with the reads to generate the next highest K value de Bruijn graph. They set the coverage for the contigs to be 1, so that the contigs in the next phase can be corrected with high coverage reads. With this approach there is a chance you could possibly add some contigs that were not connected in the reads, adding more errors, or perhaps they make the right connection giving you the correct contig. But, if there is an error, this error is propagated through each K value de

Bruijn graph, exacerbating the errors and propagating the error through multiple portions of the assembly process. For example, if there are two connected components in a higher K value graph that are connected and merged using the lower K value graph without creating any other branches, the component merging will be propagated through each subsequent K value. However, by using all of the information at the same time, it is possible that the mistake will be avoided because there will be multiple contigs that generate a branching point between the connected components and the algorithm will not merge the components together.

3   Algorithm and Implementation

Instead of iteratively building each K value graph, our approach *builds all lower value K graphs simultaneously in parallel*, then does correction on each of those graphs in parallel before generating contigs for each of the K graphs. This ensures that any error that would occur will not propagate through K different de Bruijn graphs/phases. Then the final de Bruijn graph is built with the highest value K from the reads adding in the corrected contigs from the lower value K.

If an error would be caught in the original SPAdes pipeline, it should still be caught in this pipeline, but now the algorithm is parallel. In our implementation we reuse the backbone code for assembly using reads and extra contigs. It is important to note that error correction must be performed for each K graph's contigs prior to using the contigs, as when we did not perform this step the quality degraded as K increased.

We implement the idea using SPAdes as a backbone. SPAdes is written in Python and C++. The C++ is used for most of the compute intensive operations, using Python as a wrapper to call different functions that are highly optimized for speed and memory written in C++. Given the existing code, the easiest solution was to use mpi4py [27] for multi node parallelization, as the main logic of iterating multi K values is in the Python portion of SPAdes. We note that read error correction, scaffolding and polishing have been parallelized, hence we focus on the multi-K graph creation, traversal, and contig generation phase, reusing a single node solution for read correction, scaffolding and polishing. In our evaluation we show numbers for the entire pipeline rather than just for the assembly portion, even though we are comparing the effectiveness of the assembly technique. For graph creation and traversal execution times alone, they can be seen in Fig. 3.

Using mpi4py, we take out the iterative portion of the code, instead parallelizing all the graph building, either on a single node or if available on multiple nodes. Our initial guess was that this approach would use a lot more memory, but in reality this portion takes much less memory than the error correction, so even on a single node the max RAM usage is almost identical to the original code (within 2 MB of memory usage on datasets tested).

Given that more K values lead to a better assembly, when the program is using multiple nodes, we try to maximize the usage of the resources for a given run, adding more K values if a node is idle so that during the first parallel graph building there aren't idle nodes. SPAdes will automatically decide the K values during assembly if a user does not specify specific values. If the program is run with 5 nodes, but the default SPAdes would only use 3 K values, we would increase the number of K values used, utilizing all 5 nodes to build a K graph. This allows for more utilization of the compute resources while improving the accuracy of the final assembly.

4   Evaluation

Datasets used are shown in Tables 1 and 2. Table 1 shows the datasets that were used in the GAGE-B experiments [32]. Table 2 has other datasets from NCBI where there is a published reference to compare against when using QUAST v5.0.2. Each table specifies the organism name, the type of technology used to sequence the organism, the website where the raw reads and reference genome were downloaded from, the read length in base pairs, the reference genome length in base pairs, the coverage, and library type. In Table 2, for the source dataset we show the run number from the NCBI SRA database. While we have experiments from single read libraries and mate pair libraries, we do not see a significant difference in the difference for execution time or quality when comparing to SPAdes v3.13.0 in the same scenario.

Experiments were run on a desktop computer, a single server node and on a multi-node cluster. The desktop has a 1 TB hard disk drive, 16 GB of RAM and an Intel(R) Xeon(R) CPU E5-1620 v4 @ 3.50 GHz, which has 4 cores. The single node has 750 GB of RAM, 2 NUMA cores with 20 cores each using the Intel(R) Xeon(R) CPU E5-2698 v4 @ 2.20 GHz. The multi-node setup has 244 GB hard disk drive per node, 2 eight-core sockets for a total of 16 Intel Xeon cores per node with 128 GB of memory yielding 8 GB of memory per core.

Experiments were run using the '-t' option specifying the number of threads to be used when executing SPAdes version 3.13.0. We used 16, 32, and 40 threads on the single server node, 4, 8, and 16 threads on the desktop node, and 16 threads on the multi-node. The lowest execution time among all options were used for the figures. On the single node server, SPAdes ran faster with 32 threads than 40 threads, presumably because of the parallel overhead vs. the amount of available parallel work. MULKSG used all available threads for each architecture.

There are various tools that have parallelized read error correction and scaffolding, but SPAdes only has these phases able to utilize multiple threads on a single node. To ensure that our QUAST statistics did not compare error correction techniques, but rather the assembly technique, MULKSG kept the error correction, scaffolding and polishing that SPAdes implements. This means that we do not use a multi-node parallel solution for these phases when we report our timings, and the timing that is reported includes the execution time for the entire pipeline. We would therefore expect that our parallel implementation would not scale linearly because of Amdahl's law, but would scale only for the graph building and contig generation portions. Figure 1 shows execution time on

Table 1. GAGE-B datasets

Dataset	Sequence technology	Dataset source	Reference source	Read length (bp)	Reference length (bp)	Coverage	Library
Aeromonas hydrophila	HiSeq	[1]	[2]	101	4744448	250X	Single read
Bacillus cereus	HiSeq	[3]	[4]	101	5224283	250X	Single read
Bacillus cereus	MiSeq	[5]	[4]	250	5224283	100X	Single read
Bacteroides fragilis	HiSeq	[6]	[7]	101	5373121	250X	Single read
Myobacterium abscessus	HiSeq	[8]	[9]	100	23319	115X	Single read
Myobacterium abscessus	MiSeq	[10]	[9]	251	23319	100X	Single read
Rhodobacter sphaeroides	HiSeq	[11]	[12]	101	114179	210X	Single read
Rhodobacter sphaeroides	MiSeq	[13]	[12]	251	114179	100X	Single read
Staphylococcus aureus	HiSeq	[14]	[15]	101	27041	250X	Single read
Vibrio cholerae	HiSeq	[16]	[17]	100	2961149	110X	Single read
Vibrio cholerae	MiSeq	[18]	[17]	251	2961149	100X	Single read
Xanthomonas axonopodix	HiSeq	[19]	[20]	100	4967469	250X	Single read

a single server node, and Fig. 4 shows execution times on a multi-node system. We report the graph creation and traversal execution times in Fig. 3.

As can be seen in Fig. 1, MULKSG is faster than SPAdes on the server node for 11 of the 16 datasets, very close on 3 of the datasets, and significantly slower on 2 of the datasets. On a small desktop computer with only 4 cores, MULKSG is slower than SPAdes as can be seen in Fig. 2. Memory usage is not reported here as MULKSG and SPAdes on both the single server node and the desktop experiments were within a few MB of memory usage at their peak usage. This is because the memory usage is the highest when performing the read error correction, not the graph creation or traversal.

We ran QUAST version 5.0.2 on the datasets, comparing the contigs and scaffolds generated by both SPAdes and MULKSG. We would expect the results to be very similar, as we use the same backbone for read error correction, traversal and post traversal polishing. We do in fact see in Table 3 that the results are almost identical, with there being a small amount of variation between SPAdes and MULKSG. The datasets that saw the most variation were Rhodobacter Sphaeroides MiSeq and Vibrio cholerae MiSeq. For Rhodobacter Sphaeroides, MULKSG and SPAdes had the same value for largest contig, NG50, NG75, but the largest alignment and NGA50 for MULKSG was 112704 versus 58318 for SPAdes, with MULKSG being a 93% improvement. For Vibrio cholerae, MULKSG had a max contig length of 552381 vs. SPAdes where the max contig length was 359056 (higher is better). This also affected the NGA50 and NGA75 (higher is better), with MULKSG being 33% higher when comparing scaffolds. The LGAX statistics are also lower for MULKSG (lower is better for LGAX) with other statistics being almost identical. For more statistics see Table 3.

Table 2. Other test datasets

Dataset	Sequence technology	Dataset source	Reference source	Read length (bp)	Reference length (bp)	Coverage	Library
Escherichia coli	HiSeq 2500	SRR8405059	[21]	100	5437407	80X	Mate pair
Pseudomonas aeruginosa	HiSeq X Ten	SRR8377271	[22]	150	6264404	200X	Mate pair
Salmonella enterica	MiSeq	SRR8420088	[23]	251	4951383	150X	Mate pair
Staphylococcus aureus	HiSeq 4000	SRR7748090	[24]	151	2821361	70X	Mate pair

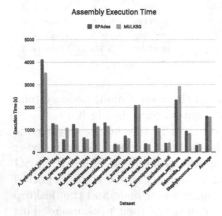

Fig. 1. Execution time for SPAdes vs. MULKSG for the entire pipeline run time when running on 1 server node.

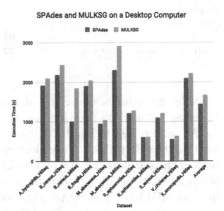

Fig. 2. Execution time for SPAdes and MULKSG for the entire pipeline run on a desktop computer.

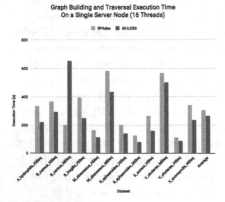

Fig. 3. Execution time for SPAdes vs. MULKSG for just the graph building and traversal time when running on 1 server node.

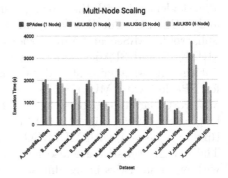

Fig. 4. Execution time for SPAdes (1 node), MULKSG (1, 2, and 6 node). Execution time includes the entire pipeline.

Table 3. Selected QUAST statistics

Assembly	Largest contig	NG50	NG75	LG50	LG75	# misassemblies	Misassembled contigs length	Genome fraction (%)	Largest alignment	NGA50	NGA75	LGA50	LGA75
A_hydrophila_HiSeq													
spades_ctg	1079605	656974	189819	3	7	6	366493	38.755	61358				
mulksg_ctg	1079605	656974	189819	3	7	6	366493	38.753	61358				
spades_scf	1079605	656974	491389	3	5	6	366493	38.682	61358				
mulksg_scf	1079605	656974	491389	3	5	6	366493	38.677	61358				
B_cereus_HiSeq													
spades_ctg	605742	136462	68074	10	25	3	4422	5.06	30865				
mulksg_ctg	605742	136462	68074	10	25	3	4403	5.059	30865				
spades_scf	638900	136462	69136	9	23	3	4422	5.065	30865				
mulksg_scf	639472	136462	69136	9	23	3	4403	5.062	30865				
B_cereus_MiSeq													
spades_ctg	676503	270560	136515	7	13	0	0	98.813	676503	270560	136515	7	13
mulksg_ctg	676503	270560	136515	7	13	0	0	98.813	676503	270560	136515	7	13
spades_scf	1202231	275327	196834	4	9	0	0	98.81	1202022	275327	196834	4	9
mulksg_scf	1202231	275327	196834	4	9	0	0	98.81	1202022	275327	196834	4	9
B_fragilis_HiSeq													
spades_ctg	639080	157629	86927	9	20	98	4146068	80.292	189113	44793	13436	35	86
mulksg_ctg	639080	157629	86927	9	20	98	4146068	80.292	189113	44793	13436	35	86
spades_scf	639080	157629	89451	9	20	99	4276309	80.292	189113	44793	13436	35	86
mulksg_scf	639080	157629	89451	9	20	99	4276309	80.292	189113	44793	13436	35	86
R_sphaeroides_HiSeq													
spades_ctg	292708	292708	292708	1	1	0	0	100	58318	58318	36015	1	2
mulksg_ctg	292708	292708	292708	1	1	0	0	100	112704	112704	112704	1	1
spades_scf	292708	292708	292708	1	1	0	0	100	58318	58318	36015	1	2
mulksg_scf	292708	292708	292708	1	1	0	0	100	112704	112704	112704	1	1
R_sphaeroides_MiSeq													
spades_ctg	465389	465389	465389	1	1	0	0	100	112703	112703	112703	1	1
mulksg_ctg	465389	465389	465389	1	1	0	0	100	112703	112703	112703	1	1
spades_scf	510758	510758	510758	1	1	0	0	100	112703	112703	112703	1	1
mulksg_scf	510758	510758	510758	1	1	0	0	100	112703	112703	112703	1	1
S_aureus_HiSeq													
spades_ctg	381434	381434	381434	1	1	0	0	31.897	4181				
mulksg_ctg	381434	381434	381434	1	1	0	0	31.897	4181				
spades_scf	381434	381434	381434	1	1	0	0	31.897	4181				
mulksg_scf	381434	381434	381434	1	1	0	0	31.897	4181				
V_cholerae_HiSeq													
spades_ctg	382053	260649	192074	5	8	3	522287	98.282	382053	192074	110108	6	11
mulksg_ctg	382053	260649	192074	5	8	2	516808	98.292	382053	192074	126528	6	11
spades_scf	382053	260649	192074	5	8	3	523776	98.286	382053	192074	110108	6	11
mulksg_scf	382053	260649	192074	5	8	2	516808	98.292	382053	192074	126528	6	11
V_cholerae_MiSeq													
spades_ctg	359056	220074	157031	6	10	1	155353	98.289	359056	215940	115075	6	11
mulksg_ctg	552381	344139	229841	4	6	1	155353	98.29	536021	246267	151655	4	8
spades_scf	384004	262473	216166	5	8	1	166802	98.284	383870	262398	115075	5	9
mulksg_scf	573239	508069	350399	3	5	1	166802	98.288	573239	350300	229773	4	6
X_axonopodis_HiSeq													
spades_ctg	312314	117399	63515	15	29	36	2455242	53.67	93507	1733		192	
mulksg_ctg	312314	117399	63515	15	29	37	2496606	53.67	93507	1820		185	
spades_scf	353048	117399	63515	14	28	36	2455313	53.67	111601	1733		191	
mulksg_scf	353048	117399	63515	14	28	37	2406677	53.67	111001	1820		184	
Escherichia_coli													
spades_ctg	373767	135151	69187	12	26	69	3861998	82.008	237170	59215	19275	25	66
mulksg_ctg	373767	135074	69187	12	26	69	3804370	82.013	237206	58620	19275	25	67
spades_scf	373767	135151	69187	12	26	69	3862256	82.018	237170	59215	19275	25	65
mulksg_scf	373767	135074	69187	12	26	69	3834628	82.017	237206	58620	19275	25	66
Pseudomonas_aeruginosa													
spades_ctg	1090886	667609	503646	4	7	40	5316041	96.306	425815	116772	70078	16	32
mulksg_ctg	1090886	667609	503646	4	7	40	5316041	96.307	425815	116772	70078	16	32
spades_scf	1090886	836250	652214	4	6	40	5316075	96.309	425815	135454	78337	14	29
mulksg_scf	1090886	836250	652214	4	6	40	5316075	96.311	425815	135454	78337	14	29
Salmonella_enterica													
spades_ctg	705817	246935	124257	6	13	53	4233225	90.489	310029	89573	41866	16	36
mulksg_ctg	705817	310911	124257	5	12	53	4232979	90.491	310029	95230	41866	15	35
spades_scf	705817	433838	167721	5	10	53	4555297	90.496	405332	107332	43685	14	32
mulksg_scf	705817	433838	167721	5	10	53	4554941	90.496	405330	107332	43685	14	32
Staphylococcus_aureus													
spades_ctg	330561	147704	77668	7	14	52	2424627	87.995	162138	70097	21795	15	34
mulksg_ctg	330561	147704	77668	7	14	52	2424697	87.996	162138	70097	21795	15	34
spades_scf	362472	157943	90405	6	12	51	2390006	87.986	162138	72762	22092	14	31
mulksg_scf	362472	157943	90405	6	12	51	2390006	87.986	162138	72762	22092	14	31

5 Related and Future Work

To our knowledge, the only multi K, multi-node de Bruijn graph assembly algorithm is ScalaDBG, which uses IDBA-UD as a backbone. ScalaDBG algorithm allows for parallel graph building, meaning the different K de Bruijn graphs are created in parallel, but then there is a serial bottleneck portion where traversal must be done on one graph (the lowest K value graph). The contigs generated from this traversal is then passed to the next highest K value de Bruijn graph, where the contigs are added in before the graph is traversed. The process continues until all K values have been used. When the contigs from one graph are added to another graph, those contigs are treated as though they have high coverage, whereas this work gives the contigs generated from a traversal a coverage of only 1, allowing for easier correction when there are errors.

IDBA-UD usually works by doing a local alignment and correction before each new K value, but for ScalaDBG this was removed because of threading dependencies that existed in IDBA-UD. This led to the need for the sequential bottleneck mentioned which was used for a type of correction to achieve the same amount of accuracy as the original IDBA-UD. When they patch existing graphs with contigs using a binary tree $log(K)$ patching approach, the accuracy was worse, with the reasoning being there was no error correction performed between patches. ScalaDBG at best has $log(K)$ patching phases, whereas MULKSG has 2 graph building/traversal phases.

We realize that our solution would not quite be used as an end to end solution at the moment, as we did not implement parallel read correction, scaffolding or polishing. We also do not do a load balancing depending on the resources for a single node, which is why with a small amount of resources on a single node our algorithm is slower than SPAdes. The number of context switches is very large when the amount of resources is scarce, hence the context switching increases the execution time. We see this is potential for future work, either modifying the SPAdes code for multi-node or taking an existing multi-node solution and substituting those chunks of functionality in the SPAdes pipeline.

6   Conclusion

To keep up with the expanding amount of data generated, bioinformatics tools are always trying to improve accuracy and speed. MULKSG is a new way to assemble a genome, clearing the sequential bottleneck that occurs in iterative multi K assembly. MULKSG breaks dependencies and allows for multi-node parallel graph creation and traversal. We implement the algorithm using SPAdes as a backbone, and show that run time can be decreased while keeping the same assembly quality, in some cases improving quality while reducing errors.

References

1. https://ccb.jhu.edu/gage_b/datasets/A_hydrophila_HiSeq.tar.gz. Accessed 7 Jan 2019
2. ftp://ftp.ncbi.nlm.nih.gov/genomes/archive/old_refseq/Bacteria/Aeromonas_hydr ophila_ATCC_7966_uid58617/NC_008570.fna. Accessed 7 Jan 2019
3. https://ccb.jhu.edu/gage_b/datasets/B_cereus_HiSeq.tar.gz. Accessed 7 Jan 2019
4. ftp://ftp.ncbi.nlm.nih.gov/genomes/archive/old_refseq/Bacteria/Bacillus_cereus_ ATCC_10987_uid57673/NC_003909.fna. Accessed 7 Jan 2019
5. https://ccb.jhu.edu/gage_b/datasets/B_cereus_MiSeq.tar.gz. Accessed 7 Jan 2019
6. https://ccb.jhu.edu/gage_b/datasets/B_fragilis_HiSeq.tar.gz. Accessed 7 Jan 2019
7. ftp://ftp.ncbi.nlm.nih.gov/genomes/archive/old_refseq/Bacteria/Bacteroides_ fragilis_638R_uid84217/NC_016776.fna. Accessed 7 Jan 2019
8. https://ccb.jhu.edu/gage_b/datasets/M_abscessus_HiSeq.tar.gz. Accessed 7 Jan 2019

9. ftp://ftp.ncbi.nlm.nih.gov/genomes/archive/old_refseq/Bacteria/Mycobacterium_abscessus_uid61613/NC_010394.fna. Accessed 7 Jan 2019
10. https://ccb.jhu.edu/gage_b/datasets/M_abscessus_MiSeq.tar.gz. Accessed 7 Jan 2019
11. https://ccb.jhu.edu/gage_b/datasets/R_sphaeroides_HiSeq.tar.gz. Accessed 7 Jan 2019
12. ftp://ftp.ncbi.nlm.nih.gov/genomes/archive/old_refseq/Bacteria/Rhodobacter_sphaeroides_2_4_1_uid57653/NC_007488.fna. Accessed 7 Jan 2019
13. https://ccb.jhu.edu/gage_b/datasets/R_sphaeroides_MiSeq.tar.gz. Accessed 7 Jan 2019
14. https://ccb.jhu.edu/gage_b/datasets/S_aureus_HiSeq.tar.gz. Accessed 7 Jan 2019
15. ftp://ftp.ncbi.nlm.nih.gov/genomes/archive/old_refseq/Bacteria/Staphylococcus_aureus_USA300_TCH1516_uid58925/NC_010063.fna. Accessed 7 Jan 2019
16. https://ccb.jhu.edu/gage_b/datasets/V_cholerae_HiSeq.tar.gz. Accessed 7 Jan 2019
17. ftp://ftp.ncbi.nlm.nih.gov/genomes/archive/old_refseq/Bacteria/Vibrio_cholerae_O1_biovar_El_Tor_N16961_uid57623/NC_002505.fna. Accessed 7 Jan 2019
18. https://ccb.jhu.edu/gage_b/datasets/V_cholerae_MiSeq.tar.gz. Accessed 7 Jan 2019
19. https://ccb.jhu.edu/gage_b/datasets/X_axonopodis_HiSeq.tar.gz. Accessed 7 Jan 2019
20. ftp://ftp.ncbi.nlm.nih.gov/genomes/archive/old_refseq/Bacteria/Xanthomonas_axonopodis_citrumelo_F1_uid73179/NC_016010.fna. Accessed 7 Jan 2019
21. ftp://ftp.ncbi.nlm.nih.gov/genomes/refseq/bacteria/Escherichia_coli/reference/GCF_000299455.1_ASM29945v1/GCF_000299455.1_ASM29945v1_genomic.fna.gz. Accessed 7 Jan 2019
22. ftp://ftp.ncbi.nlm.nih.gov/genomes/refseq/bacteria/Pseudomonas_aeruginosa/refcrence/GCF_000006765.1_ASM676v1/GCF_000006765.1_ASM676v1_genomic.fna.gz. Accessed 7 Jan 2019
23. ftp://ftp.ncbi.nlm.nih.gov/genomes/refseq/bacteria/Salmonella_enterica/reference/GCF_000006945.2_ASM694v2/GCF_000006945.2_ASM694v2_genomic.fna.gz. Accessed 7 Jan 2019
24. ftp://ftp.ncbi.nlm.nih.gov/genomes/refseq/bacteria/Staphylococcus_aureus/reference/GCF_000013425.1_ASM1342v1/GCF_000013425.1_ASM1342v1_genomic.fna.gz. Accessed 7 Jan 2019
25. Bankevich, A., et al.: SPAdes: a new genome assembly algorithm and its applications to single-cell sequencing. J. Comput. Biol. **19**(5), 455–477 (2012)
26. Batzoglou, S., et al.: ARACHNE: a whole-genome shotgun assembler. Genome Res. **12**(1), 177–89 (2002)
27. Dalcin, L.D., Paz, R.R., Kler, P.A., Cosimo, A.: Parallel distributed computing using Python. Adv. Water Resour. **34**(9), 1124–1139 (2011). https://doi.org/10.1016/j.advwatres.2011.04.013. http://www.sciencedirect.com/science/article/pii/S0309170811000777, new Computational Methods and Software Tools
28. Gurevich, A., Saveliev, V., Vyahhi, N., Tesler, G.: QUAST: quality assessment tool for genome assemblies. Bioinformatics **29**(8), 1072–1075 (2013). https://doi.org/10.1093/bioinformatics/btt086. https://www.ncbi.nlm.nih.gov/pubmed/2342 2339, 23422339[pmid]
29. Huang, X., Madan, A.: CAP3: a DNA sequence assembly program. Genome Res. **9**(9), 868–877 (1999). https://www.ncbi.nlm.nih.gov/pubmed/10508846, 10508846[pmid]

30. Li, D., Liu, C.M., Luo, R., Sadakane, K., Lam, T.W.: MEGAHIT: an ultra-fast single-node solution for large and complex metagenomics assembly via succinct de Bruijn graph. Bioinformatics **31**(10), 1674–1676 (2015). https://doi.org/10.1093/bioinformatics/btv033

31. Luo, R., et al.: SOAPdenovo2: an empirically improved memory-efficient short-read de novo assembler. Gigascience **1**(1), 18 (2012)

32. Magoc, T., et al.: GAGE-B: an evaluation of genome assemblers for bacterial organisms. Bioinformatics **29**(14), 1718–1725 (2013). https://doi.org/10.1093/bioinformatics/btt273. https://www.ncbi.nlm.nih.gov/pubmed/23665771, 23665771[pmid]

33. Mahadik, K., Wright, C., Kulkarni, M., Bagchi, S., Chaterji, S.: Scalable genomic assembly through parallel de Bruijn graph construction for multiple K-mers. In: BCB (2017)

34. Mullikin, J.C., Ning, Z.: The phusion assembler. Genome Res. **13**(1), 81–90 (2003). https://doi.org/10.1101/gr.731003. https://www.ncbi.nlm.nih.gov/pubmed/12529309, 12529309[pmid]

35. Peng, Y., Leung, H.C., Yiu, S.M., Chin, F.Y.: IDBA-UD: a de novo assembler for single-cell and metagenomic sequencing data with highly uneven depth. Bioinformatics **28**(11), 1420–1428 (2012)

36. Simpson, J.T., Wong, K., Jackman, S.D., Schein, J.E., Jones, S.J., Birol, I.: ABySS: a parallel assembler for short read sequence data. Genome Res. **19**(6), 1117–1123 (2009)

37. Zerbino, D.R., Birney, E.: Velvet: algorithms for de novo short read assembly using de Bruijn graphs. Genome Res. **18**(5), 821–829 (2008)

Counting Sorting Scenarios
and Intermediate Genomes
for the Rank Distance

João Paulo Pereira Zanetti[1] (ID), Leonid Chindelevitch[2] (ID),
and João Meidanis[1](✉) (ID)

[1] Institute of Computing, University of Campinas,
Av. Albert Einstein, 1251, Campinas, SP, Brazil
{joao.zanetti,meidanis}@ic.unicamp.br
[2] School of Computing Science, Simon Fraser University,
8888 University Drive, Burnaby, BC, Canada
leonid@sfu.ca

Abstract. An important problem in genome comparison is the genome sorting problem, that is, the problem of finding a sequence of basic operations that transforms one genome into another whose length (possibly weighted) equals the distance between them. These sequences are called optimal sorting scenarios. However, there is usually a large number of such scenarios, and a naïve algorithm is very likely to be biased towards a specific type of scenario, impairing its usefulness in real-world applications. One way to go beyond the traditional sorting algorithms is to explore all possible solutions, looking at all the optimal sorting scenarios instead of just an arbitrary one. Another related approach is to analyze all the intermediate genomes, that is, all the genomes that can occur in an optimal sorting scenario. In this paper, we show how to count the number of optimal sorting scenarios and the number of intermediate genomes between any two given genomes, under the rank distance.

Keywords: Genome rearrangements · Enumeration · Breakpoint graph

1 Introduction

At a high level of abstraction, genomes can be represented as a list of blocks (that can be genes, markers, or other syntenic regions), and their evolution can be modeled by large-scale mutation events we call *genome rearrangements*. Different genome rearrangement models define different events and costs (weights) for them, and their most basic application is to determine the lowest cost required to transform one genome into another. This is the *genome distance problem*.

JPPZ is supported by FAPESP grant 2017/02748-3. LC is supported by an NSERC Discovery Grant and a Sloan Foundation Fellowship. JM is supported by FAPESP grant 2018/00031-7.

Although this definition does not require it, in this paper we make the assumption that the genomes being considered have the same gene content. Consequently, as the rearrangement events modeled here do not alter the content of genomes, we do not model loss or duplication of blocks. Although there are models that include insertion and deletion events [3,5] and duplications [8], we do not include them here, instead focusing only on events that alter the order of the blocks.

Another fundamental problem that applies to a genome distance is to find sequences of rearrangement operations with minimum cost. We call this the *genome sorting problem*. However, a single arbitrary sequence of events among many optimal ones is hardly representative of the evolutionary process, especially considering that sorting algorithms might be biased towards certain kinds of sorting sequences. On the other hand, listing all possible optimal scenarios is not practical, because their number is simply too large.

A first step towards exploring the solution space of the genome sorting problem as a whole is to count the number of optimal sorting scenarios between two given genomes. This can reveal helpful patterns in the optimal solutions. The sorting scenario as a sequence of operations taking one genome into another is not the only way to represent a solution to the sorting problem. Shao, Lin, and Moret suggested a structure called the trajectory graph, which groups together scenarios with commuting or non-interfering operations [20].

Larget, Kadane, and Simon enumerated possible inversion scenarios in circular unichromosomal genomes in order to compute probability distributions of ancestral genome rearrangements [14]. For multi-chromosomal genomes, to the best of our knowledge, the studies of the solution space of genome sorting have been limited to the Double-Cut-and-Join (DCJ) [23] and the Single-Cut-or-Join (SCJ) [9] distances. For the latter model, Miklós, Kiss, and Tannier showed how to count optimal SCJ scenarios between two genomes [16]. In the DCJ model, the problem becomes more complex due to the possibility of recombinations. Braga and Stoye showed how to count optimal DCJ scenarios, with and without recombination [4]. Ouangraoua and Bergeron also counted optimal scenarios without recombination, and established bijections between these scenarios and known combinatorial objects [17]. Feijão took these results one step further, and showed how to count the intermediate genomes between two genomes and how to use those to reconstruct ancestral genomes [7]. Here, we produce an analogous set of results for the rank distance model [24] instead of the DCJ model.

The rank distance, introduced by Zanetti, Biller, and Meidanis [24] is based on a representation of genomes using matrices. The rank distance is twice the algebraic distance [10], and very close to the DCJ distance [10,23]. By using the rank distance, we can add linear algebra tools to the traditional breakpoint graph analysis techniques usually employed in rearrangement studies. The rank distance is equivalent to the DCJ for genomes with the same free ends. One advantage of the rank distance is that it ensures that there is no recombination while sorting. On the other hand, since the basic operations have different weights, the scenarios have variable lengths, and this represents a challenge when counting all possible scenarios. Fortunately, we were able to overcome this difficulty by adding an extra parameter in the recurrence relationship for the counts.

The rest of this paper is organized as follows. Section 2 introduces the background concepts we use: the rank distance, its basic operations, and the multi-genome breakpoint graph. In Sect. 3 we show how to count the number of optimal scenarios between two genomes in the rank distance. In Sect. 4 we count the number of intermediate genomes. In Sect. 5 we discuss our experiments on real genomic data. Finally, we summarize our work in Sect. 6. Most proofs, details, and extra material are contained in the Appendix (http://www.ic.unicamp.br/%7Emeidanis/research/rear/).

2   Background

In this section we give the theoretical background needed to explore the solution space in the next sections. First, we present the basic concepts of the rank distance model for comparing genomes. Next, we discuss the breakpoint graph and how it relates to the rank distance. Finally, we show the three basic operations necessary and sufficient for sorting genomes under the rank distance.

2.1   Genomes as Matrices and the Rank Distance

In this subsection we introduce our representation of genomes as matrices, the rank distance, and its basic operations.

We define genomes as a collection of *chromosomes*, each of them a linear or circular sequence of *genes*. A gene is a linear segment with two *extremities*: a *tail* and a *head*. When two genes appear consecutively in a chromosome, we indicate this fact by linking their closest extremities with an *adjacency*. Adjacencies are thus unordered pairs of extremities. An extremity may not be involved in more than one adjacency. If an extremity is not in any adjacency, it is called a *free end*. A genome is completely characterized by its adjacencies and free ends. Figure 1 shows an example of a genome with three genes.

We can encode this representation of a genome into a matrix [24]. In this paper we focus on comparing genomes with the same gene content. To represent genomes with matrices, we first fix an ordering of all the extremities, and use this ordering for the rows and the columns of a matrix. For instance, in Fig. 1, there are three genes, a, b, and c, and the ordering is a_t, a_h, b_t, b_h, c_t, c_h. After fixing the order of the extremities (and hence the meaning of the rows and columns of a matrix), we define the corresponding *genome matrix* as follows:

$$A_{ij} = \begin{cases} 1 & \text{if } i \neq j \text{ and } i, j \text{ are adjacent in } G, \text{ or } i = j \text{ and } i \text{ is a free end in } G \\ 0 & \text{otherwise} \end{cases}$$

An example of a genome matrix can be seen in Fig. 1. These matrices have the following properties: they are symmetric, that is, $A^T = A$, orthogonal, that is, $A^T = A^{-1}$, and binary, that is, $A \in \{0,1\}^{2n \times 2n}$, where n is the number of genes. In particular, this implies that they are involutions, i.e. permutations of order 1 or 2. These matrices are therefore sparse, and can be stored efficiently.

If A and B are genomes over the same genes, the *distance* $d(A, B)$ between A and B is defined as

$$d(A, B) = r(B - A),$$

where r is the rank of a matrix.

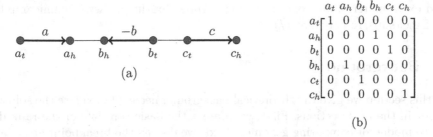

$$
\begin{array}{c|cccccc}
 & a_t & a_h & b_t & b_h & c_t & c_h \\
\hline
a_t & 1 & 0 & 0 & 0 & 0 & 0 \\
a_h & 0 & 0 & 0 & 1 & 0 & 0 \\
b_t & 0 & 0 & 0 & 0 & 1 & 0 \\
b_h & 0 & 1 & 0 & 0 & 0 & 0 \\
c_t & 0 & 0 & 1 & 0 & 0 & 0 \\
c_h & 0 & 0 & 0 & 0 & 0 & 1
\end{array}
$$

(b)

Fig. 1. (a) Genome with one linear chromosome, and adjacencies $\{a_h, b_h\}$ and $\{b_t, c_t\}$. The extremities a_t and c_h are free ends. (b) Matrix representation of the same genome.

Although the rank distance is defined in terms of matrices, it is also possible to characterize it as the weight of an optimal series of operations transforming A into B. We say that a matrix X is *applicable* to a genome A if $A + X$ is also a genome. A matrix applicable to at least one genome is called an *operation*.

An operation is *basic* if it is a *cut* of an adjacency $\{x, y\} \to \{x\}\{y\}$, a *join* of two free ends $\{x\}\{y\} \to \{x, y\}$, or a *double swap* of two adjacencies into two new ones using the same four extremities, for example $\{x, y\}\{a, b\} \to \{x, a\}\{y, b\}$. Any other rearrangement operation can be decomposed as a sum of these three kinds of operations, as we show in Sect. 2.3. We are thus able to narrow down the list of operations to just three, without loss of generality. In the context of this work, the most important information about the basic operations is that cuts and joins are rank 1 matrices, while double swaps have rank 2.

2.2   Breakpoint Graph

Genomes, as we describe here, are matchings over the set of gene extremities. We can graphically represent two genomes A and B as matchings, using one color (or line style) for A and another for B, as shown in Fig. 2.

The graph we use, called *breakpoint graph* [22] is based on the original breakpoint graph introduced by Hannenhalli and Pevzner in 1995 [11]. The difference is that we do not use caps at chromosome ends. Our free ends are simply extremities that are not adjacent to any other extremity. We believe that this makes the presentation simpler and clearer.

Given two genomes A and B of equal gene content, we build the two-genome breakpoint graph $BG(A, B)$ as follows. Its vertices are the extremities of the

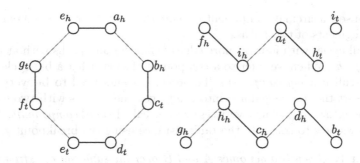

Fig. 2. Example breakpoint graph $BG(A, B)$. Genome A has adjacencies $a_h b_h$, $c_t d_t$, $e_t f_t$, $e_h g_t$, $a_t i_h$, $g_h h_h$, and $c_h d_h$, drawn as dashed edges, and free ends b_t, f_h, h_t, and i_t. Genome B has adjacencies $a_h e_h$, $b_h c_t$, $d_t e_t$, $f_t g_t$, $f_h i_h$, $a_t h_t$, $c_h h_h$, and $b_t d_h$, drawn as solid edges, and free ends g_h and i_t.

genomes, and there are two sets of edges: dashed edges, which connect pairs of extremities that are adjacent in A, and solid edges, which connect pairs of extremities that are adjacent in B.

Every node in the breakpoint graph has degree 0, 1, or 2. Because of this, the connected components of the graph are disjoint cycles and paths. The paths can be further classified into two types: paths with an even or odd number of edges. The breakpoint graph is very similar to another structure widely used in rearrangement analysis, the adjacency graph [1]. In fact, the breakpoint graph is the line graph of the adjacency graph [22]. Furthermore, as the adjacency graph has no isolated vertices, both graphs have the same number of paths and cycles.

The process of *sorting* consists of the application of basic operations to genome A until we get genome B. When sorting from A to obtain B, we call A the *source genome*, and B the *target genome*. In the graph $BG(A, B)$, we draw the edges from the source genome A with dashed lines, and the edges from the target genome B with solid edges.

Paths with an even number of edges are called *balanced*, because they have the same number of edges of each type. Odd paths are called *unbalanced*; they begin and end with edges from the same genome. Unbalanced paths can be further classified into two types, *dashed paths*, and *solid paths*, according to which genome accounts for more edges.

Note that when both genomes are equal, the edges of both colors coincide, leaving only two types of components, cycles with two edges (shared adjacencies), and isolated vertices (shared free ends). Therefore, considering $BG(A, B)$, sorting A into B can be seen as the process of applying basic operations to the dashed edges until they all coincide with the solid edges.

A cut either transforms a cycle into a path, or splits one path into two. A join does the reverse: it either transforms a path into a cycle, or joins two paths into one. A double swap can extract a cycle from any type of component or reverse a part of a component. When acting on two separate components, a double swap can also insert a circular component into another (linear or circular) one or swap the end segments of two paths.

We call cycles with four or more edges *big cycles* and paths with at least one edge *big paths*. When we refer to a component that is either a big cycle or a big path, we call it a *big component*. These components need to be worked on in order to sort the source genome into the target one. Cycles with two edges and paths of length zero (isolated vertices) will be called *small components*. The next result shows how to compute the rank distance using the breakpoint graph.

Theorem 1. *Given two genomes A and B over the same set of extremities, the rank distance between them is given by*

$$d(A, B) = 2n - 2c - p,$$

where n is the number of genes, and c and p are respectively the number of cycles and paths in $BG(A, B)$

Proof. Feijão and Meidanis in 2013 showed that the algebraic distance $d_{alg}(A, B)$ between two genomes A and B is given by $d_{alg}(A, B) = n - n_C - \frac{n_P}{2}$, where n is the number of genes, and n_C and n_P are respectively the number of cycles and paths in the adjacency graph of A and B [10, Theorem 3.11]. Since the breakpoint graph and the adjacency graph have the same number of cycles and paths, we have $c = n_C$ and $p = n_P$. Later, Zanetti, Biller and Meidanis showed that $d(A, B) = 2d_{alg}(A, B)$ [24], and, therefore, $d(A, B) = 2n - 2c - p$.   □

2.3   Sorting

In this subsection we show that is possible to sort genome A into genome B using only the three basic operations: cut, join, and double swap.

Let $\mathcal{X} = (X_1, X_2, \ldots, X_k)$ be a sequence of operations such that, for every $1 \leq i \leq k$, the operation X_i is applicable to $A + X_1 + \ldots + X_{i-1}$, and $A + X_1 + \ldots + X_k = B$. We say that $\mathcal{X}$ is a *sorting scenario* from A to B. The *weight* of $\mathcal{X}$ is the sum of the weights of its operations, that is,

$$w(\mathcal{X}) = \sum_{i=1}^{k} r(X_i).$$

We denote $w(A, B)$ the minimum weight of a sorting scenario from A to B. When a scenario $\mathcal{X}$ from A to B has $w(\mathcal{X}) = w(A, B)$, we say $\mathcal{X}$ is *optimal*. Any operation in an optimal scenario is called a *sorting operation*.

A series of lemmas leads to the desired results. We only state the main conclusion here. All the supporting proofs can be found in the appendix.

Theorem 2. *Given two genomes A and B,*

$$d(A, B) = w(A, B).$$

3   Counting the Number of Scenarios

An optimal solution for sorting genome A into genome B is a sequence of genomes separated by basic operations, going from A to B with minimum cost. Some authors call such a sequence a *geodesic* between A and B, or a *geodesic patch* when the basic operations have different weights [12]. Braga and Stoye present a similar definition for *sorting scenarios* in their work, as a sequence of operations involved in sorting A into B. We define an *operation scenario* from A to B as a list of matrices $\mathcal{L} = [X_1, \ldots, X_\ell]$ such that $A + X_1 + X_2 + \ldots + X_\ell = B$, and each matrix X_i is one of the basic operations and applicable to $A + X_1 + X_2 + \ldots + X_{i-1}$. The *total weight* of a scenario, denoted by $w(\mathcal{L})$, is the sum of the weights of all its operations. Such a scenario is *optimal* if and only if $w(\mathcal{L}) = d(A, B)$. An *optimal operation* is any basic operation that is the first in an optimal scenario. In the context of the rank distance, optimal scenarios are geodesic patches, because basic operations can have weight 1 or 2.

In this section, we show how to count the number of optimal rank sorting scenarios between two genomes, with the aid of the breakpoint graph. First, we show that no optimal operation acts on different components of the breakpoint graph. Therefore, we can solve each component separately. We then recall a formula from Braga and Stoye [4] to count DCJ sorting scenarios, which also applies to rank sorting scenarios for cycles, and a recurrence to count rank sorting scenarios for paths. Finally, we show how to get the count for the whole graph.

3.1   Recombination

As a first step to count the number of sorting scenarios, we want to prove a useful property, namely, that no optimal rearrangement recombines the components of the breakpoint graph. In other words, we show that no optimal operation acts on the extremities of more than one component at the same time.

Lemma 3. *No optimal operation in sorting from A to B involves extremities in different components of $BG(A, B)$.*

Proof. To prove this, we list every possible operation recombining two components C_1 and C_2 of $BG(A, B)$ and show that they do not change the graph in a way that reduces the distance.

A double swap can be applied to two cycles, generating one cycle; it can be applied to a cycle and a path, generating a path; or it can be applied to two paths, resulting in two paths. All of these moves reduce the number of components or keep their number unchanged, and are therefore not optimal.

The other option of a basic operation on two separate components is a join of two paths; however, this results in a single path, reducing the number of components, and is therefore also not optimal.

A cut is not considered here, because it only acts on two connected extremities. Therefore, it never affects more than one component.   □

According to Braga and Stoye, under the DCJ model the only operation that recombines different components of the graph is a double swap between two unbalanced paths resulting in two balanced ones [4]. With regard to the breakpoint graph, the difference between the DCJ and the rank distances is that, under the rank distance, every path counts towards reducing the distance, not just balanced ones. Because of this difference, recombining two unbalanced paths into two balanced ones is not an optimal move in a rank sorting scenario.

With this result, we conclude that it is possible to count the optimal scenarios for each component in the breakpoint graph independently. Now we determine how to obtain sorting scenarios for the whole graph from the separate solutions for each component.

Let s_1 and s_2 be scenarios for the components C_1 and C_2 respectively, with respective lengths ℓ_1 and ℓ_2. From Lemma 3, the number of scenarios resulting in the combination of s_1 and s_2 is the number of sequences that have both as subsequences, and this is the shuffle product of s_1 and s_2, whose size is given by the binomial coefficient $\binom{\ell_1+\ell_2}{\ell_1,\ell_2} = \frac{(\ell_1+\ell_2)!}{\ell_1!\ell_2!}$. In general, the number of sequences obtained by shuffling k subsequences is given by the multinomial coefficient $\binom{\ell_1+\ell_2+...+\ell_k}{\ell_1,\ell_2,...,\ell_k} = \frac{(\ell_1+\ell_2+...+\ell_k)!}{\ell_1!\ell_2!...\ell_k!}$, where ℓ_i is the length of the ith subsequence.

3.2 Cycles

Given a cycle in $BG(A, B)$, the only optimal operation that can be applied to it is a double swap that splits the cycle into two smaller ones, as illustrated in Fig. 3. This is a rank-two operation that increases the number of cycles by one, therefore (by Theorem 1) decreasing the distance by two.

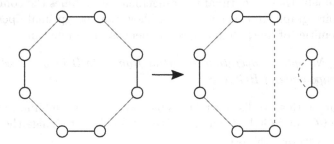

Fig. 3. One example of optimal rearrangement for a cycle. A double swap splits the cycle into two smaller ones. Here, an 8-cycle is decomposed into a 6-cycle, which will require two more double swaps to complete the sorting, and a 2-cycle, already sorted.

In this case, the sorting moves are equivalent to the ones for the DCJ distance. We can use the same formula for DCJ to compute the number $S_c(2\ell + 2)$ of scenarios to solve a cycle of length $2\ell + 2$ [4, Theorem 3]:

$$S_c(2\ell + 2) = (\ell + 1)^{(\ell-1)}$$

Each of these $S_c(2\ell+2)$ scenarios has length ℓ, and total weight 2ℓ. When A and B are co-tailed genomes (that is, A and B have the same free ends), the only big components in $BG(A, B)$ are cycles, and we can compute the total number S_{ct} of optimal sorting scenarios from A to B [4, Theorem 4] as follows:

$$S_{ct} = \frac{(\ell_1 + \ell_2 + \ldots + \ell_p)!}{\ell_1!\ell_2!\ldots\ell_p!} \prod_{i=1}^{p}(\ell_i + 1)^{\ell_i - 1},$$

where p is the number of big cycles in $BG(A, B)$.

3.3 Paths

We have shown a simple formula to compute the number of scenarios for the cyclic components in $BG(A, B)$. For paths the computation is less straightforward. The obstacle that arises when sorting paths is that, because cuts and joins have rank 1, while double swaps have rank 2, scenarios have variable length, and information on the length of the sub-solutions is necessary for the shuffling. Thus, we need a recurrence with two variables: the length of the path, and the length of the scenario.

For a path, three optimal operations are possible. First, a cut of any dashed edge, provided there is at least one dashed edge in the path. Such a cut results in two smaller paths that are solved separately. A second option is to execute a double swap on any two dashed edges, resulting in a cycle and a path, provided there are at least two dashed edges. Here again, both new components have independent scenarios that are then shuffled. The last option is to join the ends of a path, if both ends are incident to solid edges. This leads to a single cycle, that we already know how to process. These possibilities are illustrated in Fig. 4.

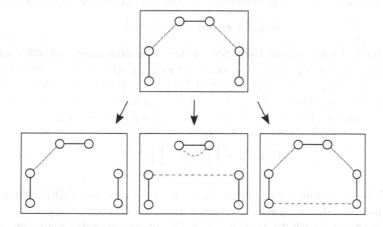

Fig. 4. Examples of the three options of optimal moves from a solid path. The first one, from the top, is a cut, splitting the 5-path into two paths with 3 and 1 edges, respectively. The second is a double swap extracting a cycle from the path. The last option, only possible in this kind of path, is to join the ends, forming a cycle.

Given a path with length k, the sizes of the optimal scenarios fall in a limited range. The longest operation scenarios for a k-path are the ones with only cuts and joins, making up a total of k operations. Since the double swaps have twice the weight of a cut or join, scenarios with more double swaps are shorter.

With this set of optimal operations, and the range for the length of a scenario, we arrive at three recurrences, one for each type of path. Since the cuts and double swaps are only applied to dashed edges, the indices are different according to the type of the path, and the paths obtained after splitting also have different types. The details are left for an appendix, for lack of space. The following theorem summarizes our results.

Theorem 4. *If the graph $BG(A, B)$ has q big paths with lengths $k_1, \ldots, k_q$, and p big cycles with lengths $2\ell_1 + 2, \ldots, 2\ell_p + 2$, the total number of optimal sorting scenarios from A to B is given by the product:*

$$Cycles(\ell_1, \ldots, \ell_p) ShufflePaths(k_1, \ldots, k_q; \ell_1, \ldots, \ell_p),$$

where $ShufflePaths(k_1, \ldots, k_q; \ell_1, \ldots, \ell_p)$ is given by

$$\sum_{\ell_1' = \lfloor k_1/2 \rfloor + 1}^{k_1} \cdots \sum_{\ell_q' = \lfloor k_q/2 \rfloor + 1}^{k_q} \frac{(\ell_1 + \ldots + \ell_p + \ell_1' + \ldots + \ell_q')!}{\ell_1! \ldots \ell_p! \ell_1'! \ldots \ell_q'!} \prod_{j=1}^{q} S_p(k_j, \ell_j')$$

and

$$Cycles(\ell_1, \ldots, \ell_p) = \prod_{i=1}^{p} S_c(2\ell_i + 2).$$

4   Intermediate Genomes

We say a genome B is an intermediate genome between genomes A and C when

$$d(A, C) = d(A, B) + d(B, C).$$

We have already shown that there is never recombination of different components in the rank distance. Therefore, we can get all possible intermediates by looking separately at each component of $BG(A, C)$. If the graph $BG(A, C)$ has p big cycles with lengths $k_1, k_2, \ldots, k_p$, and q big paths with lengths $k_1', k_2', \ldots, k_q'$, the total number $I(A, C)$ of intermediate genomes between A and C is

$$I(A, C) = \prod_{i=1}^{p} I_c(k_i) \prod_{j=1}^{q} I_p(k_j'),$$

where $I_c(k)$ is the number of intermediates for a k-cycle, and $I_p(k)$ is the number of intermediates for a k-path.

For cycles of length $2k$, with $k \geq 1$, rank optimal operations are the same as DCJ optimal operations, so we can use the known result for the DCJ distance [7]:

$$I_c(2k) = \frac{1}{k+1} \binom{2k}{k}.$$

For paths, we derive in the appendix a similar formula for $I_p(k)$, the number of intermediate genomes for a path of length $k \geq 0$:

$$I_p(k) = \binom{k+1}{\lfloor (k+1)/2 \rfloor}.$$

The following theorem summarizes the intermediate counts.

Theorem 5. *If the graph $BG(A, B)$ has p big cycles with lengths $2\ell_1, \ldots, 2\ell_p$, and q big paths with lengths $k_1, \ldots, k_q$, the total number of intermediate genomes between A and B is given by:*

$$\prod_{i=1}^{p} I_c(2\ell_i) \prod_{j=1}^{q} I_p(k_j) = \prod_{i=1}^{p} \frac{1}{\ell_i + 1} \binom{2\ell_i}{\ell_i} \prod_{j=1}^{q} \binom{k_j + 1}{\lfloor (k_j + 1)/2 \rfloor}.$$

4.1 Relationship with Sperner Families

The number $I_p(k)$ is equivalent to the number of $\lfloor (k+1)/2 \rfloor$-element subsets of a set with $(k+1)$ elements. According to Sperner's Theorem [15,21], this is the maximum number of subsets of a set with $k+1$ elements where no set contains another. Such a family of sets is called a *Sperner family*. Building on this idea, we provide a bijection between the intermediates of a k-path, and the Sperner family of all $\lfloor (k+1)/2 \rfloor$-element subsets of a $k+1$-set in the Appendix.

5 Experiments

We implemented our formulas and tested our method for counting the scenarios and the intermediates between pairs of a number of genomes from the literature.

5.1 Data Sets

We used four data sets from different sources. The first and simplest data set is from the work of Palmer and Hebron on plants of the Brassica genus [18]. It consists of two pairs of circular mitochondrial DNA, comparing *Brassica campestris* against *B. oleracea* and *B. napus*. Both instances have 5 synteny blocks. Other comparisons in this work have insertions or deletions and were not considered.

Another data set is the human and mouse X chromosome, from Pevzner and Tesler [19]. This pair of linear, single-chromosome, inputs has 11 synteny blocks.

The third data set is composed of 13 chloroplast genomes, of which 12 are from the Campanulaceae family, and 1 from tobacco as an outgroup, with 105 synteny blocks, taken from Cosner, Raubeson and Jansen [6], and also used by Bourque and Pevzner [2]. These created 78 pairs of inputs to the sorting problem.

The fourth and largest data set contains genomes used by Kim et al. to test their reconstruction algorithm DESCHRAMBLER [13]. It consists of 20 instances comparing the human genome against the genome from other animals, namely, 18 Eutherian mammals, plus opossum and chicken as outgroups. These pairs have between 101 and 621 synteny blocks.

5.2 Code

The code to run our experiments was implemented in Python, and executed on a virtual machine using a single 2.3 GHz processor core and 2 GB of memory. Building the breakpoint graph of our tests instances, with up to 1242 extremities, is not a computationally intensive task, and neither is computing the number of intermediates, a simple product of binomial coefficients. The most demanding task is to compute the number of scenarios, especially when the breakpoint graph has a large number of paths. This is a consequence of the fact that each path adds an extra summation in the expression for the number of scenarios.

5.3 Results and Discussion

The two Brassica instances have the same rank distance of 6 (in this case three double swaps), and the same results: 9 scenarios and 10 intermediate genomes.

For the pair of X chromosomes, with a rank distance of 14, we get 237440 scenarios and 560 intermediates.

With the chloroplast genome pairs, we get varying results. Some pairs, like *Trachelium* and *Campanula* are only one double swap apart, and therefore have only one optimal scenario and two intermediates (the input genomes). The pair of genomes that are farthest apart is *Merciera* and *Platycodon*, with a distance of 48. They have 1.4×10^{32} optimal scenarios, and 4.9×10^{12} intermediate genomes.

With the Eutherian data set, due to the large number of terms in the summation in Theorem 4 for the number of scenarios between distantly related species, we developed a scheme that is always guaranteed to use a specified amount of memory, but sometimes ends up producing upper and lower bounds on the number of scenarios instead of the exact values. This scheme produced the exact result for 5 out of the 20 instances, namely, the primates (chimpanzee, marmoset, orangutan, and rhesus) and the horse. For the other instances, we obtained upper and lower bounds on the number of optimal scenarios, using the monotonicity of $S_p(k, l)$ and multinomial coefficients containing l with respect to the l variable, and utilizing up to $N = 10^9$ memory cells, as described in the Appendix. On the other hand, computing the number of intermediates proved easy even for the farthest pairs of genomes. In Table 1 we list the number of scenarios (or an interval containing it) and the number of intermediates for all the instances.

Comparing the larger instances of the chloroplast data set with the ones from the human, cat and mouse genomes, we note that linear, multichromosomal genomes tend to have more scenarios than circular, unichromosomal ones. This may be due to the fact that each path in the breakpoint graph adds another summation in the scenario formula, consisting of many products. In contrast, each cycle only adds one product to existing terms. The number of intermediates, on the other hand, is likely to be less variable between instances with the same distance, since the formulas for intermediates in paths and cycles are very similar.

Table 1. Rank distance, number of scenarios, and number of intermediates between the human genome and the genomes of 20 Eutherian animals, listed in order of distance.

Genome	d	Scenarios	Intermediates
Chimpanzee	27	6.54×10^{11}	2.46×10^4
Orangutan	53	6.03×10^{38}	1.29×10^{10}
Rhesus	150	1.21×10^{138}	1.45×10^{28}
Marmoset	204	3.99×10^{250}	3.13×10^{43}
Horse	225	1.63×10^{135}	1.31×10^{51}
Dog	304	$[10^{432}, 10^{471}]$	6.37×10^{70}
Pig	318	$[10^{463}, 10^{479}]$	1.61×10^{73}
White rhino	328	$[10^{546}, 10^{588}]$	1.36×10^{84}
Elephant	336	$[10^{583}, 10^{609}]$	8.56×10^{86}
Cattle	383	$[10^{537}, 10^{579}]$	7.39×10^{83}
Pika	385	$[10^{647}, 10^{710}]$	2.89×10^{98}
Goat	393	$[10^{548}, 10^{588}]$	5.30×10^{85}
Tenrec	407	$[10^{700}, 10^{778}]$	1.71×10^{105}
Shrew	487	$[10^{876}, 10^{999}]$	1.02×10^{128}
Mouse	509	$[10^{830}, 10^{980}]$	2.44×10^{131}
Manatee	519	$[10^{1050}, 10^{1101}]$	3.12×10^{142}
Guinea pig	640	$[10^{1130}, 10^{1360}]$	5.85×10^{166}
Chicken	736	$[10^{1217}, 10^{1550}]$	6.38×10^{193}
Opossum	778	$[10^{1295}, 10^{1590}]$	4.01×10^{204}
Rat	788	$[10^{1251}, 10^{1476}]$	2.30×10^{189}

6   Conclusion

In this paper we opened the doors for the exploration of the solution space of the rank distance problem. We demonstrated that there is no recombination between components of the breakpoint graph in any optimal rank sorting scenario. We then gave a formula for the number of optimal sorting scenarios for co-tailed genomes, and presented a general algorithm for counting the number of scenarios.

We also presented a formula for the number of intermediates between two genomes. In addition, we constructed a bijection that provides a simple way to uniformly sample intermediates. Sampling intermediate genomes is the next step in the study of the solution space and can be very helpful in future applications.

References

1. Bergeron, A., Mixtacki, J., Stoye, J.: A unifying view of genome rearrangements. In: Bücher, P., Moret, B.M.E. (eds.) WABI 2006. LNCS, vol. 4175, pp. 163–173. Springer, Heidelberg (2006). https://doi.org/10.1007/11851561_16
2. Bourque, G., Pevzner, P.A.: Genome-scale evolution: reconstructing gene orders in the ancestral species. Genome Res. **12**(1), 26–36 (2002)
3. Braga, M.D.V., Willing, E., Stoye, J.: Genomic distance with DCJ and indels. In: Moulton, V., Singh, M. (eds.) WABI 2010. LNCS, vol. 6293, pp. 90–101. Springer, Heidelberg (2010). https://doi.org/10.1007/978-3-642-15294-8_8
4. Braga, M.D., Stoye, J.: The solution space of sorting by DCJ. J. Comput. Biol. **17**(9), 1145–1165 (2010)
5. Compeau, P.E.C.: DCJ-Indel sorting revisited. Algorithms Mol. Biol. **8**(1), 6 (2013)
6. Cosner, M.E., Raubeson, L.A., Jansen, R.K.: Chloroplast DNA rearrangements in Campanulaceae: phylogenetic utility of highly rearranged genomes. BMC Evol. Biol. **4**(1), 1–17 (2004)
7. Feijão, P.: Reconstruction of ancestral gene orders using intermediate genomes. BMC Bioinform. **16**(14), S3 (2015)
8. Feijão, P., Mane, A., Chauve, C.: A tractable variant of the single cut or join distance with duplicated genes. In: Meidanis, J., Nakhleh, L. (eds.) RECOMB-CG 2017. LNCS, vol. 10562, pp. 14–30. Springer, Cham (2017). https://doi.org/10.1007/978-3-319-67979-2_2
9. Feijao, P., Meidanis, J.: SCJ: a breakpoint-like distance that simplifies several rearrangement problems. IEEE/ACM Trans. Comput. Biol. Bioinform. (TCBB) **8**(5), 1318–1329 (2011)
10. Feijão, P., Meidanis, J.: Extending the algebraic formalism for genome rearrangements to include linear chromosomes. IEEE/ACM Trans. Comput. Biol. Bioinform. **10**(4), 819–831 (2013)
11. Hannenhalli, S., Pevzner, P.A.: Transforming men into mice (polynomial algorithm for genomic distance problem). In: 1995 Proceedings of the 36th Annual Symposium on Foundations of Computer Science, pp. 581–592. IEEE (1995)
12. Jamshidpey, A., Jamshidpey, A., Sankoff, D.: Sets of medians in the non-geodesic pseudometric space of unsigned genomes with breakpoints. BMC Genomics **15**(6), S3 (2014)
13. Kim, J., et al.: Reconstruction and evolutionary history of eutherian chromosomes. Proc. Nat. Acad. Sci. **114**(27), E5379–E5388 (2017)
14. Larget, B., Kadane, J.B., Simon, D.L.: A Bayesian approach to the estimation of ancestral genome arrangements. Mol. Phylogenet. Evol. **36**(2), 214–223 (2005)
15. Lubell, D.: A short proof of Sperner's lemma. J. Comb. Theory **1**(2), 299 (1966)
16. Miklós, I., Kiss, S.Z., Tannier, E.: Counting and sampling SCJ small parsimony solutions. Theor. Comput. Sci. **552**, 83–98 (2014)
17. Ouangraoua, A., Bergeron, A.: Combinatorial structure of genome rearrangements scenarios. J. Comput. Biol. **17**(9), 1129–1144 (2010)
18. Palmer, J.D., Herbon, L.A.: Plant mitochondrial DNA evolved rapidly in structure, but slowly in sequence. J. Mol. Evol. **28**(1), 87–97 (1988)
19. Pevzner, P., Tesler, G.: Genome rearrangements in mammalian evolution: lessons from human and mouse genomes. Genome Res. **13**(1), 37–45 (2003)
20. Shao, M., Lin, Y., Moret, B.: Sorting genomes with rearrangements and segmental duplications through trajectory graphs. BMC Bioinform. **14**(15), S9 (2013)

21. Sperner, E.: Ein satz über untermengen einer endlichen menge. Math. Z. **27**(1), 544–548 (1928)
22. Tannier, E., Zheng, C., Sankoff, D.: Multichromosomal median and halving problems under different genomic distances. BMC Bioinform. **10**(1), 120 (2009)
23. Yancopoulos, S., Attie, O., Friedberg, R.: Efficient sorting of genomic permutations by translocation, inversion and block interchange. Bioinformatics **21**(16), 3340–3346 (2005)
24. Zanetti, J.P.P., Biller, P., Meidanis, J.: Median approximations for genomes modeled as matrices. Bull. Math. Biol. **78**(4), 786–814 (2016)

Generalizations of the Genomic Rank
Distance to Indels

João Paulo Pereira Zanetti[1] (ID), Leonid Chindelevitch[2] (ID),
and João Meidanis[1(✉)] (ID)

[1] Institute of Computing, University of Campinas, Av. Albert Einstein,
1251, Campinas, SP, Brazil
{joao.zanetti,meidanis}@ic.unicamp.br
[2] School of Computing Science, Simon Fraser University, 8888 University Drive,
Burnaby, BC, Canada
leonid@sfu.ca

Abstract. The rank distance model, introduced by Zanetti *et al.* in
2016, represents genome rearrangements in multi-chromosomal genomes
looking at them as matrices. So far, this model only supported comparisons between genomes with the same gene content. We seek to generalize
it, allowing for genomes with different gene content. In this paper, we
approach such generalization from two different angles, both using the
same representation of genomes, and leading to simple distance formulas and sorting algorithms for genomes with different gene contents, but
without duplications.

Keywords: Genome rearrangements · Substitutions ·
Breakpoint graph

1 Introduction

In the context of genome comparison, one can view a genome as a collection of
contiguous, conserved segments arranged in linear and/or circular chromosomes.
These segments can be genes or more general markers. Using this abstraction, we
pay no attention to smaller mutations affecting just a few nucleotides, and focus
instead on bigger mutations that move larger portions of the genome, changing
the order of segments with respect to one another. We call these bigger mutations
genome rearrangements.

In simpler models of genome rearrangement, the operations only move
genomic segments around, without creating or destroying markers. However, to
better reflect genome evolution, it is desirable to include operations that alter the
content of the genome. For example, we may consider operations that add contiguous segments to the genome, called *insertions*, and operations that remove

JPPZ is supported by FAPESP grant 2017/02748-3. LC is supported by an NSERC
Discovery Grant and a Sloan Foundation Fellowship. JM is supported by FAPESP
grant 2018/00031-7.

© Springer Nature Switzerland AG 2019
I. Holmes et al. (Eds.): AlCoB 2019, LNBI 11488, pp. 152–164, 2019.
https://doi.org/10.1007/978-3-030-18174-1_11

contiguous segments from the genome, called *deletions*. In general, we call these two types of operation *indels*.

To the best of our knowledge, the work on including indels in genome rearrangement models has so far been limited to the inversion distance [8] for unichromosomal genomes, and the Double-Cut-and-Join (DCJ) distance [14] on multichromosomal genomes [3,11].

In 2001, El-Mabrouk first studied the problem of sorting by inversions and indels, developing an exact algorithm for the cases where there were only insertions or deletions, but not both [6]. In 2008, Yancopoulos and Friedberg proposed extending the DCJ model to account for insertions and deletions [15], and in 2010 Braga et al. presented a linear time algorithm for the DCJ-Indel problem [4]. Later, Compeau used a different approach, looking at indels as DCJ operations themselves, and arrived at a simpler DCJ-Indel distance formula and sorting algorithm [5]. Another extension of the DCJ model by Braga et al. comes from adding a more powerful operation: a *substitution* of a genome segment for another [2]. The development of DCJ-Indel also led to advancements on the inversion-indel distance, by Willing et al. in 2013 [13].

In this paper, we explore the addition of indels to the rank distance model, which was initially developed for same-content genomes [16]. In this model, genomes are represented as matrices, and the distance between two genomes is the rank of their difference. We expect this model to have a natural extension to genomes with unequal content, leading to simple formulas and algorithms.

The rest of this paper is organized as follows. Section 2 presents the background on the rank distance and defines the representation of genomes that do not necessarily have all the markers being considered. In Sect. 3 we expand the rank distance to encompass genomes with different genomic content. In Sect. 4 we present a different approach for adding indels to the rank distance model. Section 5 describes our experiments, and Sect. 6 presents our conclusions. Most proofs, details, and extra material are contained in the Appendix (http://www.ic.unicamp.br/%7Emeidanis/research/rear/).

2   Definitions

2.1   Markers, Genomes, and Matrices

We begin our definitions with the notion of a *marker*, which is a contiguous DNA stretch that is conserved in all genomes where it appears. This will be our building block in constructing genomes.

Let $\mathcal{G}$ be a set of markers. Each marker $g \in \mathcal{G}$ has two extremities: a head g_h, and a tail g_t. The set

$$V(\mathcal{G}) = \{g_h, g_t | g \in \mathcal{G}\}$$

contains all extremities associated to $\mathcal{G}$. We will fix a 1–1 mapping identifying $V(\mathcal{G})$ with the canonical basis $\{e_1, e_2, \ldots, e_{2n}\}$ of $\mathbb{R}^{2n}$, where $n = |\mathcal{G}|$ and e_i is the $2n \times 1$ column vector whose i^{th} entry is 1 and all others are 0. Since this mapping is fixed, we will use the same letter and type font to denote both an extremity x and its corresponding column vector.

A genome A over $\mathcal{G}$ consists of a set $V(A) \subseteq V(\mathcal{G})$ of extremities and a set $E(A)$ of *adjacencies*, which are unordered pairs of distinct extremities from $V(A)$, with the extra restriction that each extremity in $V(A)$ can belong to at most one adjacency. Note that a genome does not necessarily contain all the extremities from all the markers in $\mathcal{G}$. We do not even require that $g_t \in V(A)$ if $g_h \in V(A)$, and vice versa. The reason for that will become clear in Sect. 3.

If a pair $\{x, y\}$ belongs to $E(A)$, we say that x and y are *adjacent* in genome A. From the definitions, we see that each extremity $x \in V(A)$ has to either be adjacent to exactly one other extremity, or be a *free end*, that is, an extremity not adjacent to any other (e.g. near the end of a linear chromosome). In addition, extremities from $V(\mathcal{G})$ that do not belong to $V(A)$ will be called A-*null*, because they will correspond to null rows and columns in the matrix for A, as we will see shortly.

For example, let $\mathcal{G} = \{a, b, c, d\}$, and let A be a genome with $V(A) = \{a_h, b_h, d_h, a_t, b_t, d_t\}$, and $E(A) = \{\{a_h, b_t\}, \{b_h, d_h\}\}$. Genome A is illustrated in Fig. 1.

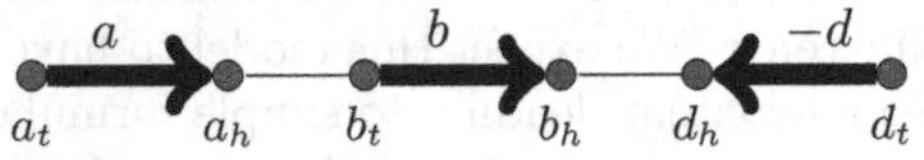

Fig. 1. Genome A with $V(A) = \{a_h, b_h, d_h, a_t, b_t, d_t\}$, and $E(A) = \{\{a_h, b_t\}, \{b_h, d_h\}\}$.

Given that extremities are identified with column vectors of $\mathbb{R}^{2n}$, we may view genomes as matrices as follows. Using the same letter and typeface A to represent the matrix associated to the genome A, we will define:

$$Ax = \begin{cases} y, & \text{when } \{x, y\} \in E(A), \\ x, & \text{when } x \text{ is a free end in } V(A), \\ 0, & \text{when } x \notin V(A). \end{cases}$$

This formula unambiguously define A, since it specifies the image under A of a basis of $\mathbb{R}^{2n}$. As an example, the matrix representation for the genome A in Fig. 1 is:

$$\begin{array}{c} \\ a_t \\ a_h \\ b_t \\ b_h \\ c_t \\ c_h \\ d_t \\ d_h \end{array} \begin{array}{c} a_t\ a_h\ b_t\ b_h\ c_t\ c_h\ d_t\ d_h \\ \left[\begin{array}{cccccccc} 1 & 0 & 0 & 0 & 0 & 0 & 0 & 0 \\ 0 & 0 & 1 & 0 & 0 & 0 & 0 & 0 \\ 0 & 1 & 0 & 0 & 0 & 0 & 0 & 0 \\ 0 & 0 & 0 & 0 & 0 & 0 & 0 & 1 \\ 0 & 0 & 0 & 0 & 0 & 0 & 0 & 0 \\ 0 & 0 & 0 & 0 & 0 & 0 & 0 & 0 \\ 0 & 0 & 0 & 0 & 0 & 0 & 1 & 0 \\ 0 & 0 & 0 & 1 & 0 & 0 & 0 & 0 \end{array}\right] \end{array},$$

where $a_t = e_1$, $a_h = e_2$, ..., $d_h = e_8$. A matrix that can be obtained from a genome in this fashion will be called a *genomic matrix*. It is easy to see that a square binary matrix $A \in \{0,1\}^{2n \times 2n}$ is genomic if and only if $A^T = A$ and A^2 is a diagonal matrix with 0's and 1's on the diagonal; indeed, the 1 entries on the diagonal of A^2 correspond to the extremities present in $V(A)$.

2.2   Rank Distance

Let A and B be two genomic matrices. We can define a distance between them as follows:

$$d_r(A, B) = r(B - A),$$

where $r(X)$ denotes the rank of matrix X. For invertible genome matrices A and B, which do not have zero rows or columns and therefore include all the extremities, this definition generalizes the rank distance of Zanetti et al. [16]. This distance satisfies the required properties for a metric:

- $d_r(A, B) = 0 \iff A = B$
- $d_r(A, B) = d_r(B, A)$
- $d_r(A, C) \le d_r(A, B) + d_r(B, C)$

For example, consider the genome A defined above, and let B be the following genome, illustrated in Fig. 2:

$$B = \begin{bmatrix} 0 & 0 & 0 & 0 & 0 & 0 & 0 & 0 \\ 0 & 0 & 0 & 0 & 0 & 0 & 0 & 0 \\ 0 & 0 & 1 & 0 & 0 & 0 & 0 & 0 \\ 0 & 0 & 0 & 0 & 1 & 0 & 0 & 0 \\ 0 & 0 & 0 & 1 & 0 & 0 & 0 & 0 \\ 0 & 0 & 0 & 0 & 0 & 0 & 1 & 0 \\ 0 & 0 & 0 & 0 & 0 & 1 & 0 & 0 \\ 0 & 0 & 0 & 0 & 0 & 0 & 0 & 1 \end{bmatrix}$$

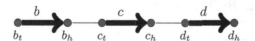

Fig. 2. Matrix and chromosomal representations of a genome B with $V(B) = \{b_h, c_h, d_h, b_t, c_t, d_t\}$, and adjacencies $\{\{b_h, c_t\}, \{c_h, d_t\}\}$.

Having matrices for both A and B on hand, we can compute their difference:

$$B - A = \begin{bmatrix} -1 & 0 & 0 & 0 & 0 & 0 & 0 & 0 \\ 0 & 0 & -1 & 0 & 0 & 0 & 0 & 0 \\ 0 & -1 & 1 & 0 & 0 & 0 & 0 & 0 \\ 0 & 0 & 0 & 0 & 1 & 0 & 0 & -1 \\ 0 & 0 & 0 & 1 & 0 & 0 & 0 & 0 \\ 0 & 0 & 0 & 0 & 0 & 0 & 1 & 0 \\ 0 & 0 & 0 & 0 & 0 & 1 & -1 & 0 \\ 0 & 0 & 0 & -1 & 0 & 0 & 0 & 1 \end{bmatrix}$$

And, finally, we have the distance $d_r(A, B) = r(B - A) = 8$. However, computing the rank of the matrix $B - A$ directly is not the most computationally efficient way to compute the rank distance. In Sect. 3.1 we will see how to do that in $O(n)$ time.

2.3   Augmented Breakpoint Graph

To prepare for the addition of indels to the rank distance model, we defined genomes so that they do not necessarily have the same gene content. We use a structure called the *augmented breakpoint graph*, analogous to the regular breakpoint graph, but, following Compeau [5], with labels at the ends of each path.

The nodes of the augmented breakpoint graph $BG(A, B)$ of A and B are the extremities of the set $V(G) \supseteq V(A) \cup V(B)$, and two nodes x and y are adjacent in $BG(A, B)$ if they are adjacent in either A or B. As in the regular breakpoint graph, all components are either paths or cycles. Sometimes we refer to them as a k-path or a k-cycle when we want to emphasize that k is the number of edges in the path or the cycle.

In the augmented breakpoint graph, all nodes with degree 2 are necessarily in the intersection $V(A) \cap V(B)$, because they are parts of adjacencies in both genomes. On the other hand, nodes x with degree 1 are path endpoints, and at least one of the following cases applies:

- x is a free end in A: $Ax = x$,
- x is a free end in B: $Bx = x$,
- x is A-null: $Ax = 0$,
- x is B-null: $Bx = 0$.

When a path has at least one edge, then it has exactly two distinct end nodes. For each of these two nodes at the ends of the path, exactly one of the cases above apply. When both endpoints are free ends, we call the path *proper*. We say a path is *A-null* (*B-null*) when one of its ends is a free end, and the other is an A-null (B-null) node. When a path has two distinct A-null (B-null) ends, we call the path *AA-null* (*BB-null*). In the case where one end is A-null and the other is B-null, the path is called *AB-null*.

Finally, when a node x has degree zero in $BG(A, B)$, exactly two of the previous cases apply, leading to four possibilities:

- When x is a free end in both A and B, it forms a proper path;
- When x is a free end in A and B-null, it forms a B-null path;
- When x is A-null and a free end in B, it forms an A-null path;
- Finally, when x is null in both A and B, the "natural" definition would be to consider it an AB-null path. However, as we will see in Sect. 4, for the rank-indel distance it makes more sense to consider this path a proper path. For the rank distance, it makes no difference to consider it as either a proper path or as an AB-null path. We choose to adopt the convention that it is a proper path, to accommodate both versions of the distance.

As an example, Fig. 3 is the augmented breakpoint graph $BG(A, B)$ of the genomes A and B seen earlier.

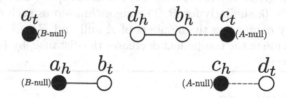

Fig. 3. Augmented breakpoint graph $BG(A, B)$. Black solid edges are adjacencies from A, gray dashed edges are from B. White nodes are extremities in both $V(A)$ and $V(B)$. Black nodes are either A-null or B-null, as specified besides them. The components are two A-null paths and two B-null paths.

Given two genomes A and B, we will define some statistics for $BG(A, B)$. We will use $c(A, B)$ and $p(A, B)$ to denote, respectively, the number of cycles and paths in $BG(A, B)$. The number of paths is the sum of the number of paths of each type: $p_0(A, B)$ is the number of proper paths in $BG(A, B)$, while $p_A(A, B)$, $p_B(A, B)$, $p_{AA}(A, B)$, $p_{BB}(A, B)$ and $p_{AB}(A, B)$ are the number of A-null, B-null, AA-null, BB-null and AB-null paths, respectively.

3 Rank Distance in the Presence of Indels

In this section we discuss the rank distance of genomes with possibly different marker content. First, in Sect. 3.1, we provide an efficient algorithm to compute the rank distance. Then, in Sect. 3.2, we define the most concise set of operations needed to transform one genome into another. Finally, in Sect. 3.3, we show how to use these operations to optimally sort genomes.

3.1 Efficient Computation of the Rank Distance

Given two genomes A and B, we prove the following theorem in the appendix:

Theorem 1.

$$d_r(A, B) = 2n - 2c(A, B) - p_0(A, B) - p_{AB}(A, B).$$

Algorithm 1 (page 7) implements these ideas and runs in $O(n)$ time, efficiently computing $d_r(A, B) = r(A - B)$. It is a Breadth-First Search traversing $BG(A, B)$ that in addition computes a score s for each component, equal to the difference between the number of A-null and B-null extremities in it. Notice that extremities i such that $A[i] > 0$ and $B[i] > 0$ contribute zero to the score. A score of zero means the component has the same number of A-null and B-null extremities, so we decrease d by 1 for a path, or by 2 for a cycle. Since the initial value of d is $2n$, we end up with $d = d_r(A, B)$.

Algorithm 1. Algorithm to compute the distance between genomes A and B. Genome A is given as a list of length $2n$, where $A[i] = j$ if $Ae_i = e_j$, and $A[i] = 0$ if $Ae_i = 0$; similarly for B. The algorithm scores each component in $BG(A, B)$ by comparing the numbers of A-null and B-null extremities. Equal numbers mean the component decreases the distance, by 1 for a path, or by 2 for a cycle.

```
d ← |A|
while ∃x not visited do
    Q ← {x}
    s ← 0
    mark x as visited
    while Q ≠ ∅ do
        take i from Q
        if both A[i] and B[i] are > 0, ≠ i, and visited then
            d ← d − 1
        if A[i] > 0 then
            s ← s + 1
            if A[i] not visited then
                mark A[i] as visited
                add A[i] to Q
        if B[i] > 0 then
            s ← s − 1
            if B[i] not visited then
                mark B[i] as visited
                add B[i] to Q
    if s = 0 then
        d ← d − 1
return d
```

3.2 Basic Operations

A matrix X is an *operation* when there is a genome A such that $A + X$ is a genome. In this case we say that X is *applicable* to A. The *weight* of an operation X is the rank of X.

When the genomes considered have the same marker content, only three types of operations are needed to sort any genome into another: cuts, joins and double swaps [10]. Cuts and joins have weight 1, while double swaps have weight 2. In this paper, we seek to add to our model operations that deal with unequal gene content. This turns out to be a highly non-trivial task, as we explain in more detail in the appendix. Here we summarize our final model.

As expected, we need to consider the insertion or the deletion of an entire chromosome. Insertions and deletions of parts of chromosomes are not needed, as we show in the appendix. The matrix for the insertion or deletion of a chromosome with k markers is, up to the sign, equivalent to a genome with k markers, and always has weight $2k$. Therefore, the weight of such an operation is $2k$.

In addition, we consider a new kind of operation that takes advantage of our relaxed definition of genomes. Recall that, when we defined genomes in Sect. 2, we mentioned that, given a genome A, we do not require that $g_t \in V(A)$ if $g_h \in V(A)$, or vice-versa. This relaxed definition now comes into play. We define an operation that substitutes a single extremity for an extremity that does not exist in the genome; due to its rank, we assign such an operation a weight of 2.

Introducing this kind of operation implies that the concept of chromosomes also has to be relaxed. In a genome where, for every $g \in \mathcal{G}$, the extremities g_h and g_t are either both present or both absent, a chromosome is a sequence of markers that can be either circular, having no free ends, or linear, with exactly two free ends. In the case of a genome with only one extremity of a marker, there are *semi-chromosomes* that, instead of ending at a free end, end with an unpaired extremity, that is, a head extremity whose corresponding tail is not in the genome, or vice versa. As a result, now an insertion or a deletion can be of a whole chromosome, or of a whole semi-chromosome, always with a weight equal to the number of extremities being inserted or deleted.

With the introduction of extremity substitutions, we now have six types of basic operations:

- Cut, with cost 1.
- Join, with cost 1.
- Double swap, with cost 2.
- Deletion of whole chromosomes or semi-chromosomes, costing the number of extremities deleted.
- Insertion of whole chromosomes or semi-chromosomes, costing the number of extremities inserted.
- Substitution of one extremity, with cost 2.

From here to the end of Sect. 3, we call these the *basic operations* to transform one genome into another when they do not share the same set of markers. We will see how the basic operations suffice to carry out such transformations.

3.3   Sorting

Let $\mathcal{X} = (X_1, X_2, \ldots, X_k)$ be a sequence of operations such that, for every $1 \leq i \leq k$, the operation X_i is applicable to $A + X_1 + \ldots + X_{i-1}$, and $A + X_1 + \ldots + X_k = B$. We say that $\mathcal{X}$ is a *sorting scenario* from A to B. The *weight* of $\mathcal{X}$ is the sum of the weights of its operations, that is,

$$w(\mathcal{X}) = \sum_{i=1}^{k} r(X_i).$$

We denote by $w(A, B)$ the minimum weight of a sorting scenario from A to B. When a scenario $\mathcal{X}$ from A to B satisfies $w(\mathcal{X}) = w(A, B)$, we call $\mathcal{X}$ *optimal*.

In the appendix we show that the rank distance $d(A, B)$ is equal to the optimum weight of a scenario going from A to B using the basic operations described in Sect. 3.2. The main result proved there is the following theorem:

Theorem 2. *Given two genomes A and B,*

$$d_r(A, B) = w(A, B).$$

Triangle Inequality. One of our main concerns with the addition of indel operations to a genomic distance is respecting the triangle inequality. When indels have a constant cost, the triangle inequality is easily violated. Consider the three genomes in Fig. 4. Only one deletion is needed to transform either A or B into C, but rearranging A into B takes more operations.

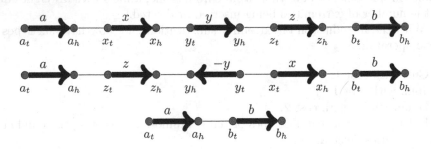

Fig. 4. Example of genomes A, B, C that can violate the triangle inequality.

Yancopoulos and Friedberg call this violation "the free lunch problem" [15]. Their suggestion to deal with this problem is to add a surcharge to the cost of an indel, based on the adjacency graph.

Braga et al. dealt with the violation of the triangle inequality by adding a simpler surcharge after the computation of their DCJ-indel distance [4]. The surcharge to the distance between A and B is equal to $ku(A, B)$, where k is a constant, and $u(A, B)$ is the number of unique markers between genomes A and B. At first they showed the triangle inequality holds when $k \geq 3/2$, but later work showed that $k \geq 1$ is a tight bound [1].

Braga et al. also defined a framework to assign variable costs to indels, a linear function of the number of markers inserted or deleted, and showed that it is equivalent to the *a posteriori* surcharge [1].

Unlike the DCJ distance, the rank distance, with the proposed extension to the matrix representation of genomes, naturally offers an indel mechanism with weights that avoid the free lunch problem.

4   An Alternative: The Rank-Indel Distance

In order to avoid general extremity substitutions and genomes without both extremities of a marker, a different approach to the addition of indels to the rank distance is to define a genomic distance that includes the basic operations of the rank distance for genomes with the same content, plus insertions and deletions, all with the same weight as in the rank distance model. This way, we define the *rank-indel distance* $d_i(A, B)$ of A and B as the minimum cost of an operation sequence sorting A into B, using the basic operations cited above:

- Cuts/joins, with cost 1.
- Double swaps, with cost 2.
- Insertions/deletions of linear or circular chromosomes with k markers, costing $2k$.

We already know that $d_i(A, B) \geq r(B - A)$. This inequality can sometimes be strict. In fact, we prove the following theorem in the Appendix:

Theorem 3. $d_i(A, B) = 2n - 2c(A, B) - p_0(A, B) + p_{AB}(A, B)$.

5   Experiments

We performed an experiment to assess how much evolutionary signal the rank and rank-indel distances are capable of capturing. We used fungal genomes from 13 species: the causal agent of rice blast, *Magnaporthe oryzae*, plus 12 isolates of *Magnaporthe grisea*, a related pathogen, collected from infected wheat plants. The data was kindly made available by A. Nhani Jr, N. Talbot, and D. Soanes.

The 12 isolates were sequenced, assembled, annotated, and mapped onto the complete genome of *M. oryzae*. The annotated genes of *M. oryzae* then provided a gene set containing all the genes in the annotated genomes of the isolates, and serve as our comprehensive marker set $\mathcal{G}$ in this analysis. The resulting .gff files were used to determine gene position and orientation of each gene in each genome. We realize that this procedure leaves out genes unique to the isolates, not present in *M. oryzae*, but, even with this bias, the results were encouraging.

Each `.gff` file contains genes assembled in 7 chromosomes, plus an extra chromosome numbered "8" where small pieces that could not be placed anywhere else during assembly were kept. In our analysis, we decided to only consider the first 7 chromosomes, as they correspond to genes that could be effectively placed onto a real chromosome. In addition, the mapping process produces a similarity coefficient, varying from 0.0 to 1.0, that tells how similar the isolate and *M. oryzae* genes are. In our analysis, we only kept the genes for which this coefficient is 1.0, meaning very high similarity. With this, we lost about 7% or all the genes. Finally, we filtered out any coding sequence (CDS) properly contained in a larger CDS, as these are likely annotation artifacts.

With the position and orientation of all genes, we built the adjacencies in each genome and input the results to our rank and rank-indel distance calculating algorithms. Table 1 contains the pairwise distances obtained. It is interesting to observe that for all pairs, the rank distance coincided with the rank-indel distance, *meaning that there are no AB-paths between any pair.* We built a phylogenetic tree with these distances using the neighbor-joining method [12], as implemented in the Mega software package [9].

Table 1. Pairwise rank and rank-indel distances (they are equal) between 13 fungal genomes. *M. oryzae* is denoted by M; other isolates by their numeric ID. Notice that the distances involving M are much larger, consistent with the fact that it infects rice while all others infect wheat.

	5033	0925	36	5003	6047	86	25	35	6017	5010	M	6045	5035
5033	0	239	332	285	378	429	361	344	185	436	13111	299	170
0925	239	0	277	323	414	444	342	307	264	417	13083	270	245
36	332	277	0	451	282	553	285	126	357	508	13126	211	316
5003	285	323	451	0	369	566	436	413	305	549	13167	419	273
6047	378	414	282	369	0	649	307	230	370	618	13215	322	348
86	429	444	553	566	649	0	499	587	452	321	13038	538	461
25	361	342	285	436	307	499	0	313	338	418	13093	242	333
35	344	307	126	413	230	587	313	0	375	558	13150	229	326
6017	185	264	357	305	370	452	338	375	0	463	13108	272	143
5010	436	417	508	549	618	321	418	558	463	0	13101	483	460
M	13111	13083	13126	13167	13215	13038	13093	13150	13108	13101	0	13115	13111
6045	299	270	211	419	322	538	242	229	272	483	13115	0	249
5035	170	245	316	273	348	461	333	326	143	460	13111	249	0

A larger set of blast pathogens was analysed by Gladieux *et al.* in a recent article [7], where phylogenetic trees were constructed based on several factors, including maximum likelihood on almost 3,000 orthologous CDS, maximum likelihood on 9 loci, and pairwise BLAST distances between repeat-masked genomes. Restricting the trees to just the common genomes, we verify that our trees only differ in the positioning of the PY6045-PY36 subtree (Fig. 5).

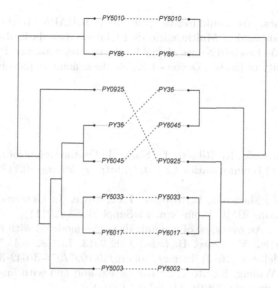

Fig. 5. Phylogenetic trees for the same genome set constructed with our distances and neighbor-joining (left) and with more traditional methods (right). Notice that the trees only differ by the placement of the PY6045-PY36 subtree.

6   Conclusion

In this paper, we expanded the rank distance to account for genomes with different gene content, but still without duplications. The first step, in Sect. 2, was to define genomes that do not necessarily contain all markers of $\mathcal{G}$. This allows for the representation of genomes with different markers from each other, and is done very naturally, by using zeros in the rows/columns corresponding to the missing markers. We then developed ways to compare these genomes.

The first approach simply extends the rank distance, keeping the distance $d_r(A, B)$ between two genomes A and B equal to the rank $r(A - B)$ of their difference. We showed how to efficiently compute d_r, and how to transform A into B using only basic operations, adding insertions, deletions, and the substitution of a single extremity to the cast of basic operations of the rank distance of genome with the same markers.

The substitution of single extremities leads to genomes with incomplete markers. To avoid this, we also present an alternative rank-indel distance that changes the content of a genome only through insertions and deletions of chromosomes. We note that both distances have very simple formulas, and are closely related, with $d_i(A, B) = d_r(A, B) + 2p_{AB}(A, B)$.

Phylogenetic trees constructed with our distances turn out to be very close to trees built with other, more traditional methods, showing that there is enough phylogenetic signal in the order and orientation of genes alone. Further studies will be conducted to better assess the usefulness of these distances in phylogenetics.

Acknowledgments. We would like to thank the EMBRAPA Multiuser Bioinformatics Laboratory (Laboratório Multiusuário de Bioinformática da Embrapa) for helpful discussions, Dr. João Leodato Nunes Maciel for the isolates data, and Prof. Dr. Nicholas J. Talbot (University of Exeter, Devon - UK) for the genome sequencing and assembly infrastructure.

References

1. Braga, M., Machado, R., Ribeiro, L., Stoye, J.: On the weight of indels in genomic distances. BMC Bioinformatics **12** (2011). https://doi.org/10.1186/1471-2105-12-S9-S13

2. Braga, M.D.V., Machado, R., Ribeiro, L.C., Stoye, J.: Genomic distance under gene substitutions. BMC Bioinform. **12**(Suppl 9), S8 (2011)

3. Braga, M.D.V.: An overview of genomic distances modeled with indels. In: Bonizzoni, P., Brattka, V., Löwe, B. (eds.) CiE 2013. LNCS, vol. 7921, pp. 22–31. Springer, Heidelberg (2013). https://doi.org/10.1007/978-3-642-39053-1_3

4. Braga, M.D., Willing, E., Stoye, J.: Double cut and join with insertions and deletions. J. Comput. Biol. **18**(9), 1167–1184 (2011)

5. Compeau, P.E.C.: DCJ-indel sorting revisited. Algorithms Mol. Biol. **8**(1), 6 (2013). https://doi.org/10.1186/1748-7188-8-6

6. El-Mabrouk, N.: Sorting signed permutations by reversals and insertions/deletions of contiguous segments. J. Discrete Algorithms **1**(1), 105–122 (2001)

7. Gladieux, P., et al.: Gene flow between divergent cereal- and grass-specific lineages of the rice blast fungus Magnaporthe oryzae. mBio **9**(1) (2018). https://doi.org/10.1128/mBio.01219-17

8. Hannenhalli, S., Pevzner, P.A.: Transforming cabbage into turnip: polynomial algorithm for sorting signed permutations by reversals. J. ACM **46**(1), 1–27 (1999). https://doi.org/10.1145/300515.300516

9. Kumar, S., Stecher, G., Li, M., Knyaz, C., Tamura, K.: MEGA X: molecular evolutionary genetics analysis across computing platforms. Mol. Biol. Evol. **35**(6), 1547–1549 (2018)

10. Meidanis, J., Biller, P., Zanetti, J.P.P.: A matrix-based theory for genome rearrangements. Technical report IC-17-11, Institute of Computing, University of Campinas, in English, 45 p., August 2017

11. Paten, B., Zerbino, D.R., Hickey, G., Haussler, D.: A unifying model of genome evolution under parsimony. BMC Bioinform. **15**(1), 206 (2014)

12. Saitou, N., Nei, M.: The neighbor-joining method: a new method for reconstructing phylogenetic trees. Mol. Biol. Evol. **4**(4), 406–425 (1987)

13. Willing, E., Zaccaria, S., Braga, M.D., Stoye, J.: On the inversion-indel distance. In: BMC Bioinformatics, vol. 14, p. S3. BioMed Central (2013)

14. Yancopoulos, S., Attie, O., Friedberg, R.: Efficient sorting of genomic permutations by translocation, inversion and block interchange. Bioinformatics **21**(16), 3340–3346 (2005)

15. Yancopoulos, S., Friedberg, R.: DCJ path formulation for genome transformations which include insertions, deletions, and duplications. J. Comput. Biol. **16**(10), 1311–1338 (2009)

16. Zanetti, J.P.P., Biller, P., Meidanis, J.: Median approximations for genomes modeled as matrices. Bull. Math. Biol. **78**(4), 786–814 (2016)

Sequence Analysis, Phylogenetics and Other Biological Processes

Using INC Within Divide-and-Conquer Phylogeny Estimation

Thien Le[1] , Aaron Sy[2], Erin K. Molloy[1] , Qiuyi (Richard) Zhang[3],

Satish Rao[2], and Tandy Warnow[1(✉)]

[1] Department of Computer Science, University of Illinois at Urbana-Champaign,
201 S. Goodwin Ave, Urbana, IL 61801, USA
{thienle2,emolloy2,warnow}@illinois.edu
[2] Department of EECS, University of California at Berkeley,
Berkeley, CA 94720, USA
{raaronsy,satishr}@berkeley.edu
[3] Department of Mathematics, University of California at Berkeley,
Berkeley, CA 94720, USA
qiuyi@gmail.com

Abstract. In a recent paper (Zhang, Rao, and Warnow, Algorithms for Molecular Biology 2019), the INC (incremental tree building) algorithm was presented and proven to be absolute fast converging under standard sequence evolution models. A variant of INC which allows a set of disjoint constraint trees to be provided and then uses INC to merge the constraint trees was also presented (i.e., Constrained INC). We report on a study evaluating INC on a range of simulated datasets, and show that it has very poor accuracy in comparison to standard methods. We also explore the design space for divide-and-conquer strategies for phylogeny estimation that use Constrained INC, and show modifications that provide improved accuracy. In particular, we present INC-ML, a divide-and-conquer approach to maximum likelihood (ML) estimation that comes close to the leading ML heuristics in terms of accuracy, and is more accurate than the current best distance-based methods.

Keywords: Inferring the evolutionary phylogeny of species ·
Phylogeny estimation · Maximum likelihood · Sample complexity ·
Divide-and-conquer

1 Introduction

Phylogeny estimation is a basic part of many biological analysis pipelines, and is generally formulated as a statistical inference problem in which the sequences

Supported by the University of Illinois at Urbana-Champaign and NSF grants DGE-1144245, CCF-1535977, and CCF-1535989. Computational experiments were performed on Blue Waters, supported by NSF grants OCI-0725070 and ACI-1238993 and by the State of Illinois.

© Springer Nature Switzerland AG 2019
I. Holmes et al. (Eds.): AlCoB 2019, LNBI 11488, pp. 167–178, 2019.
https://doi.org/10.1007/978-3-030-18174-1_12

given as input are assumed to have been generated on an unknown (but fixed) model tree, equipped with a stochastic model of evolution. One of the interesting questions in phylogeny estimation is whether a method is guaranteed to converge to the model tree as the sequence length goes to infinity; methods that have this property are said to be "statistically consistent".

Although statistical consistency is important, the "sample complexity" (which evaluates the amount of data that a method needs to return the true tree with high probability) is perhaps more important than statistical consistency, since datasets are of finite length. We describe this issue in the context of the well known Jukes-Cantor model [8], and say that the phylogeny estimation method Φ is *absolute fast converging (AFC)* for the Jukes-Cantor (JC) model if there is a polynomial $p(n)$ such that for all ϵ, f, g the probability of recovering the true tree T given sequences of length $p(n)$ is at least $1 - \epsilon$ for all JC model trees (T, Θ) where T has n leaves and all edges e satisfy $f \leq l_e \leq g$ (where l_e is the expected number of times a random site will change on edge e). The first provably AFC methods were presented in [5,6], and were distance-based methods that operated by computing quartet trees and then combining them. By restricting the quartet trees that were used to a subset (called the "short quartets") of the full set of quartet trees, it was shown that the true tree could be constructed with high probability from polynomial length sequences. Since then, many other methods have been established to be AFC, including maximum likelihood [22].

Recently, a new polynomial time AFC method, INC, was presented by Zhang *et al.* [29]. Zhang *et al.* also presented a variant of the method called INC-NJ that uses NJ on small subsets, and proved that INC-NJ is AFC and has low degree polynomial time. Finally, Zhang *et al.* presented a generic technique called constrained-INC that allows the user to provide a set of disjoint constraint trees, and then uses INC to combine the constraint trees into a tree on the full dataset. However, no implementation of INC was developed, and so INC and its variants were not explored with respect to empirical accuracy on data.

Our study presents an extensive study of INC and its variants, as described in [29], and also explores modifications to INC to improve accuracy. We explore the design space of divide-and-conquer strategies using INC, and compare the best of these methods to standard phylogeny estimation methods on simulated data. Finally, we conclude with a discussion of future work. Our open-source implementation, as well as all commands necessary to reproduce the study, are available at [10]. All datasets generated for this study are available at [11].

2 The Incremental Tree Building (INC) Method

The input to INC is a set of sequences S in an alignment $\mathcal{A}$, and a distance matrix d computed on the alignment. An ordering on the sequences is computed from the input matrix d: first a minimum spanning tree is computed, and the taxa (i.e., sequences) are ordered so that each added sequence is adjacent within the spanning tree to one sequence that precedes it in the ordering. The taxa are added into a growing tree t, which begins with the first three taxa in the ordering and then adds each new taxon into t. To add a new taxon q into t, a set of quartet trees is computed, and these quartet trees vote on where to place q into t. The approach in [29] uses the Four Point Method (FPM) [3,4] to compute quartet trees and then allows only "valid quartets" (which are quartets with sufficiently low interleaf distances, according to the input matrix d) to vote. Furthermore, all quartet trees have the same voting power. When all the taxa have been inserted, the final tree is returned. Because of the incremental nature of the approach, the algorithm is called Incremental Tree Building, or INC.

We compare INC and INC-NJ, both of which are AFC, to two ML heuristics (FastTree2 [21] and RAxML [24]) and two distance-based methods (BME, balanced minimum evolution, within FastME [12] and Neighbor Joining within PAUP* [25]). Maximum likelihood, if solved exactly, is also absolute fast converging [22], but it is unlikely that the two heuristics are AFC. We explore their relative performance in terms of FN error rate (i.e., missing branch rate), which is the fraction of the bipartitions defined by internal branches in the true tree that are missing from the estimated tree, on 20 datasets with 1000 sequences that evolve under a high rate of evolution (the 1000L1 model from [16]). FastTree2 and RAxML are run with default settings under the Generalized Time Reversible (GTR) model of sequence evolution [26], BME (balanced minimum evolution) is run within FastME2 [12] using NNI and SPR moves, and NJ is run in default settings within PAUP*, both given logdet distance matrices computed by PAUP*.

All analyses were limited to 48 h on Blue Waters (a supercomputer at the University of Illinois); RAxML was run on multiple threads, but even so some RAxML analyses did not complete within that time. We report the best ML solution found within the 48 h time limit in such cases.

Table 1. Tree error rates on 20 replicates of the 1000L1 model condition.

Method	INC	INC-NJ	FastTree2	NJ	FastME	RAxML
Tree error rate	0.910	0.707	0.109	0.434	0.307	0.117

INC and INC-NJ both have very high error rates of 91% and 70.7%, respectively, and the other methods have much lower error rates (Table 1). The best accuracy is obtained using the two maximum likelihood heuristics (error rates under 12%), and the two distance-based methods have moderate error rates (43.4% for NJ and 30.7% for FastME). Thus, INC is much less accurate than INC-NJ and NJ is much better than INC-NJ. Since NJ is *not* AFC [9] and both INC and INC-NJ are AFC, this result is disappointing (to say the least). However, this experiment does not address whether using ML heuristics to compute constraint trees would result in improved accuracy compared to other methods. Exploring this question is the purpose of the next section.

3 Exploring the Design Space for Constrained-INC

Zhang et al. [29] showed that INC can be used to combine a given set of disjoint trees (i.e., constraint trees). This approach, called Constrained-INC, can be used within a divide-and-conquer strategy for tree construction: the input set of sequences is divided into disjoint subsets, trees are constructed on subsets (using a preferred method), and then the subset trees are combined using INC into a tree on the full dataset.

Prior divide-and-conquer strategies [27] have been helpful in terms of improving scalability of phylogeny estimation methods to large datasets [1,20,28], but have produced trees on overlapping subsets, which cannot be analyzed using constrained-INC. Instead, we use the centroid edge decomposition strategy in SATé-II [13] and PASTA [16] (methods for co-estimating alignments and trees) that operates as follows: an initial tree is computed, then the tree is decomposed into two subtrees of (roughly) the same size (by pulling out a "centroid edge"), and the process recurses until all the subsets are small enough.

We explore this design space, evaluating the impact of the following algorithmic parameters: (a) initial tree and decomposition size, (b) how constraint trees are computed, (c) how quartet trees (for merging constraint trees) are computed, and (d) the weight of the votes (i.e., identical weights for all quartets, or weights that depend on the specific quartet tree). Details about how we generated the datasets and performed the analyses are available at the github site for INC.

We use four different collections of simulated datasets ranging from 101 taxa to 10,000 taxa, using the missing branch error rate. The first experiment explores the design space of constrained INC on 20 replicates of the 1000L1 datasets. We include a comparison to INC-NJ, which uses a specific decomposition strategy and then computes constraint trees using neighbor joining (NJ), as described in [29]. The result of this experiment produced two divide-and-conquer strategies that use INC to combine constraint trees. In our second experiment, we compare these two strategies to NJ, RAxML, and FastTree2.

Datasets. We analyzed several collections of sequence alignments from different model conditions that varied in terms of number of sequences, sequence length, degree of deviation from a clock, the substitution model, branch lengths, and whether or not the sequence evolution model has insertions and deletions (only the 1000L1 datasets have indels; all the others are gap-free). The datasets we explore vary in difficulty, with the most difficult datasets having high rates of evolution, as reflected in the average and maximum p-distances (i.e., normalized Hamming distances, see Table 2).

101-taxon datasets. The model tree has very short internal branches, reflecting a rapid radiation, and evolve under the GTR model without indels (20 replicates). These are not ultrametric. The sequence datasets were generated for [23].

500-taxon Kuhner-Felsenstein (KF) datasets. The model trees for these datasets were produced by a random process developed by Kuhner and Felsenstein, designed to produce tree topologies and branch lengths similar to those seen in biological datasets. The initial tree taken from [2] has average branch length of 1, and we rescaled it to produce varying rates of evolution. These trees are far from ultrametric. Sequences evolve under the Jukes-Cantor model [8].

1000L1 datasets. We use the true alignments from 20 replicates from the 1000L1 datasets studied in [16]. These datasets have a high rate of evolution with both substitutions and long insertions and deletions (indels), with average gap length 13.6, and do not have short branches. The model tree is not ultrametric. The average percentage of the true alignment that is gapped (i.e., gappiness) is 73.2%.

1000-taxon (SB) datasets. These date were generated by taking a single model gene tree from [19] with very short branches (SB), representing a rapid radiation, and is not ultrametric. We then varied the rate of evolution by rescaling branch lengths, and generated 20 replicate datasets for each rescaled model tree using INDELible [7] with the same GTR+gamma model parameters as in [19].

10K datasets. This model tree has 10,000 leaves and the topology and branch lengths have a strict molecular clock with have many short branches. We generated this model tree by first generating a species tree using SimPhy [15] (with parameters consistent with a rapid radiation, similar to the 1000-taxon short branch datasets) and then evolving a single gene within the species tree. Finally, we varied branch lengths by rescaling the gene tree and evolved sequences under the GTR+gamma model using INDELible; 10 replicate datasets were generated for each scaling factor.

Table 2. Average and maximum p-distances (max = 1.0) are given for each model condition.

Datasets	1000L1	500	101	1000(SB)	10K
Average p-distance	0.70	0.72	0.13	0.21	0.19
Maximum p-distance	0.77	0.81	0.33	0.32	0.30

3.1   Results for Experiment 1

In our first experiment, we compare divide-and-conquer strategies that use INC. We use the centroid edge decomposition (but changing the target subset size) to define the subsets, construct trees on the subsets using different techniques, and then combine the subset (constraint) trees using different ways of running constrained-INC, including specific changes to its algorithm design. We explore these variants on the 20 replicates of the 1000L1 datasets. Unless specified otherwise, we use the following settings for all methods: INC-ML uses FastTree2 as a starting tree, uses a centroid edge decomposition, constructs the constraint trees using FastTree under the GTR model, and employs unweighted voting restricted to the valid quartets. Neighbor joining (NJ) is run using PAUP*, BME is run using FastME2 [12] with NNI and SPR searches, and FastTree2 and RAxML are run in default mode under the GTR model.

Impact of subset size. We began by evaluating the impact of size of the subsets produced by the centroid edge decomposition strategy, using FastTree2 to compute the constraint trees on the disjoint subsets. As shown in Table 3, the largest subsets produce the best accuracy. Since the largest subsets also have the highest running time, we explore results using maximum subset size of 200 for the remaining studies in Experiment 1.

Table 3. Results for Experiment 1: impact of maximum subset size on tree error rates (maximum 1.0) for INC-ML. Results shown are mean error rates over 20 replicates of the 1000L1 model condition.

Maximum subset size	20	50	100	200	500
Tree error rate for variants of INC-ML	0.70	0.50	0.37	0.26	0.17

Table 4. Results for Experiment 1: table showing impact on final tree error rate for INC-ML of the choice of starting tree (maximum 1.0); results shown are averages over 20 replicates of the 1000L1 model condition.

Starting tree	FastTree2	FastMe	NJ
Tree error rate for variants of INC-ML	0.261	0.272	0.277

Impact of the choice of the initial tree. Next we study the impact of the choice of the initial tree. The result of this experiment (Table 4) shows that using FastTree2 to compute the initial tree gives the overall best accuracy; hence, we use FastTree2 for the initial tree in our remaining experiments.

Impact of voting scheme. The original voting scheme, presented in [29] allows only valid quartets to vote, and each vote has the same weight. Given the new taxon q to add and growing tree t_g, each internal node i defines a quartet $Q_i = \{u_1, u_2, u_3, q\}$ and all valid quartets (as defined in [29]) have unit weight. Furthermore, given a valid quartet tree $u_1u_2|u_3q$, this identifies a subtree of the growing tree into which q can be added. The quartet tree adds one vote to each edge in that subtree. The implementation of this voting scheme is performed with a straightforward breadth-first-search [29].

We modify this strategy by changing which quartets are allowed to vote and redefining the weight of their votes; we also consider schemes that have two phases. Since increases in the diameter of a quartet (which is defined by the distance matrix d) are known to increase the error rate in the estimated quartet tree [4], we consider weighting schemes that depend on the diameter of the quartet. Overall, we evaluate the following five protocols.

Voting protocol 1 (VP1): The valid quartets are used without weights to identify a set of edges that have the maximum total support. If there is more than one edge with maximum total support, the first edge encountered with that maximum support is selected.

Voting protocol 2 (VP2): This protocol uses two phases. The first phase is as with the first protocol in that the valid quartets without weights vote to identify a set E^* of edges that have the maximum total support. Then, the valid quartets are used with weights to select from among the set E^* of identified edges. If there is still a tie, the first edge encountered with that maximum support is selected. The weight of a quartet tree with diameter (maximum length between any 2 leaves) d_Q is set to either $\frac{1}{d_Q}$ (VP2.1) or $\frac{1}{d_Q^2}$ (VP2.2).

Voting protocol 3 (VP3): This protocol uses one phase. All valid quartets vote with weights (see below), and the set of edges that have maximum total support is identified. If there is a tie, the first edge encountered with that maximum support is selected. The weight of a quartet tree with diameter d_Q is set to $\frac{1}{d_Q^2}$.

Voting protocol 4 (VP4): This protocol uses two phases. In the first phase, all valid quartets vote with weights (see below), and the set E^* of edges that have maximum total support is identified. If there is a tie, then all quartets are allowed to vote (but only on the set E^*) with weights; this produces a subset of E^* that has the maximum total support. If there is still a tie, then the first edge encountered with that maximum support is selected. The weight of a quartet tree with diameter d_Q is set to $\frac{1}{d_Q^2}$ in both rounds.

Voting protocol 5 (VP5): This protocol uses one phase. All quartets vote with weights (see below) and the set of edges that have maximum total support is identified. If there is a tie, the first edge encountered with that maximum support is selected. The weight of a quartet tree with diameter d_Q is set to $\frac{1}{d_Q^2}$.

As shown in Table 5, VP3-VP5 produced slightly better accuracy than VP1, VP2.1, and VP2.2. Of the three better voting schemes, VP3 and VP5 have the advantage of using only one phase, and VP3 has a (slight) running time advantage over VP5 in that it only allows valid quartets to vote. Therefore, in subsequent analyses we used VP3.

Table 5. Results for Experiment 1: the impact of voting schemes on tree error rates (means over 20 replicates) for INC-ML on the 1000L1 datasets.

Voting scheme	VP1	VP2.1	VP2.2	VP3	VP4	VP5
Tree error rate for variants of INC-ML	0.266	0.269	0.269	0.249	0.249	0.249

Impact of how constraint trees are computed. We examine three ways of computing constraint trees: RAxML used to estimate the constraint tree for each subset, FastTree2 used to estimate the constraint tree for each subset, and the induced tree on the specified subset of the FastTree2 tree on the full set of taxa. As shown in Table 6, using the induced trees on each subset computed by Fast-Tree2 (i.e., FastTree2-induced) produced the best accuracy; hence, we use this approach in subsequent studies.

Table 6. Results for Experiment 1: the impact on tree error rates of how constraint trees are computed: RAxML, FastTree2, or the induced subtree of the FastTree2 tree on the full set of taxa (i.e., "FastTree2-induced"). Results shown are averages over 20 replicate datasets from the 1000L1 model condition.

Constraint tree method	FastTree2	FastTree2-induced	RAxML
Tree error rates for variants of INC-ML	0.249	0.247	0.255

Impact of how quartet trees are computed. We then explored how the quartet trees are computed. We explored three techniques: the Four Point Method (FPM, the default in [29]), RAxML on each quartet, and using the induced quartet tree from the FastTree2 tree on the full dataset. The best result was obtained using the induced tree from FastTree2 (Table 7). Hence, for subsequent experiments, we compute quartet trees by restricting the FastTree2 tree to each quartet.

Table 7. Results for Experiment 1: impact of methods on INC-ML of methods used to compute quartet trees.

Quartet tree method	FPM	FastTree2-induced	FastTree2
Tree error rate for variants of INC-ML	0.247	0.109	0.137

Summary of Experiment 1. Experiment 1 showed that changes to the Constrained-INC design could result in improved accuracy, with some algorithmic parameters having large impacts. In particular: the size of the constraint trees was important (with larger subsets better), how quartet trees are computed was important (with the Four Point Method much less accurate than FastTree2-induced quartet trees), and other parameters providing a small improvement. In Experiment 2, we maintained the settings selected in Experiment 1.

3.2   Results for Experiment 2

In our second experiment, we explore two variants of INC-ML: one (INC-ML (fast)) that is designed for speed and the other (INC-ML (slow)) that is slower and designed for improved accuracy. Both variants use the same divide-and-conquer strategy, differing only in the ML heuristic they use to construct trees on subsets (RAxML for the slow variant and FastTree2 for the fast variant). Each method uses FastTree2 to compute an initial tree, divides the dataset into subsets with at most $n/5$ taxa (where n is the number of taxa in the input set) using the centroid edge decomposition, and constructs ML trees on each subset. Each method uses induced quartet trees from the FastTree2 starting tree for the quartet trees (only on the valid quartets) and voting scheme VP3.

Overall results for the basic model conditions are shown in Table 8. We also show results as we scale the branch lengths in each model tree for the 10K-taxon

Table 8. Tree error rates of two variants of INC-ML, two ML heuristics, and NJ under different model conditions (over 10 replicates for the 10K-taxon condition and 10 replicates for each other model condition).

Datasets	101	1000L1	1000(SB)	10K
INC-ML (Fast)	0.453	0.266	0.170	0.130
INC-ML (Slow)	0.453	0.253	0.212	0.155
FastTree2	0.453	0.109	0.180	0.131
NJ	0.500	0.434	0.421	0.271
RAxML	0.375	0.117	0.123	0.096

datasets in Fig. 1; results for the 1000-taxon datasets with short branches show the same trends and are omitted due to space constraints. Results are quite consistent: both variants of INC-ML are more accurate than NJ, come close to FastTree2 in terms of accuracy (and are sometimes more accurate), but are generally less accurate than RAxML.

Table 9. Average runtime (seconds) over 10 replicates of 10K model condition and 20 replicates on the other model conditions. The asterisk (*) indicates that RAxML could not complete on any of the 10 replicates within 48 h on the 10K datasets.

Datasets	101	1000L1	1000(SB)	10K
INC-ML (Fast)	26	376	182	4182
INC-ML (Slow)	29	1121	750	48385
FastTree2	7	233	75	4071
NJ	0	2	3	2212
RAxML	32	4187	2827	(*)172800

Runtimes are shown in Table 9. Unsurprisingly, NJ is the fastest of the methods, completing in less time than the other methods, and FastTree2 is the next fastest method. The slowest method by far is RAxML, which does not complete on any of the 10K datasets within 48 h. Finally, as expected, INC-ML (Slow) is slower than INC-ML (Fast), reflecting that the slow version uses RAxML and the fast version uses FastTree2.

3.3   Comparison Between Constrained-INC and NJMerge

To the best of our knowledge, the only other method that combines disjoint trees is NJMerge [18,19], which was explored for multi-locus species tree estimation in which gene tree heterogeneity due to ILS is present [14]. As shown in [18,19], NJMerge provided substantial advantages for coalescent-based species tree estimation, and in particular it maintained accuracy but reduced running time for

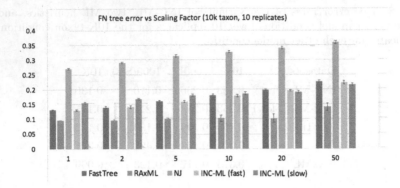

Fig. 1. Tree error rates on the 10K datasets for two variants of INC-ML, two ML heuristics, and NJ, as the model tree is scaled from a low rate of evolution (scaling factor 1) to a high rate of evolution (scaling factor 50).

ASTRAL [17], which is the current leading coalescent-based method. However, NJMerge was not explored in the context of GTR phylogeny estimation, where there has been several decades of substantial effort in developing highly accurate maximum likelihood codes. We compare constrained-INC and NJMerge on the same 1000-taxon datasets studied in this paper, using the same constraint trees, to evaluate the impact of choice of merger technique (i.e., NJMerge or INC). As shown in Table 10, the two methods have essentially identical accuracy.

Table 10. Tree error rates on the 1000-taxa short branch datasets for INC-ML and NJMerge [18] on RAxML constraint trees, as a function of the scaling factor.

Scaling factor	0.2	0.5	1	2	5	10	20	50
Tree error rate for INC-ML (slow)	0.16	0.18	0.21	0.19	0.20	0.22	0.23	0.26
Tree error rate for NJMerge (slow)	0.16	0.18	0.19	0.20	0.20	0.22	0.23	0.26

4   Conclusions

This study has shown that the original design for INC and INC-NJ, despite the excellent theoretical guarantee of being absolute fast converging (AFC) under standard sequence evolution models, does not provide very good accuracy compared to even simple distance-based methods. However, the specifics of the algorithm design mattered: changing how quartet trees are computed and allowing quartet trees to vote improved accuracy. These modifications in algorithm design enabled INC-ML to produce trees that came close to the accuracy of the best maximum likelihood methods and that were more accurate than the leading distance-based methods.

Despite the somewhat disappointing performance of INC-ML, the results shown here are limited to just one divide-and-conquer strategy, and it is possible that other divide-and-conquer strategies might provide better accuracy. However, it seems likely that the best that divide-and-conquer approaches can provide is a *speed-up* rather than an improvement in accuracy over RAxML, which is regarded as one of the best maximum likelihood heuristics.

Finally, the similarity in performance between INC and NJMerge strongly suggests that INC should be useful in other settings, such as multi-locus species tree estimation, where this type of divide-and-conquer strategy does provide improvements in running time—without sacrificing accuracy. Further research is needed to explore the design space and the different application domains for divide-and-conquer strategies.

References

1. Bayzid, M.S., Hunt, T., Warnow, T.: Disk covering methods improve phylogenomic analyses. BMC Genomics **15**(Suppl 6), S7 (2014)
2. Boc, A., Diallo, A., Makarenkov, V.: T-REX: a web server for inferring, validating and visualizing phylogenetic trees and networks. Nucleic Acids Res. **40**, W573–W579 (2012)
3. Buneman, P.: A note on the metric properties of trees. J. Comb. Theory (B) **17**, 48–50 (1974)
4. Erdős, P., Steel, M., Székely, L., Warnow, T.: Local quartet splits of a binary tree infer all quartet splits via one dyadic inference rule. Comput. Artif. Intell. **16**(2), 217–227 (1997)
5. Erdős, P., Steel, M., Székely, L., Warnow, T.: A few logs suffice to build (almost) all trees (I). Random Struct. Algorithms **14**, 153–184 (1999)
6. Erdős, P., Steel, M., Székely, L., Warnow, T.: A few logs suffice to build (almost) all trees (II). Theor. Comput. Sci. **221**, 77–118 (1999)
7. Fletcher, W., Yang, Z.: INDELible: a flexible simulator of biological sequence evolution. Mol. Biol. Evol. **26**(8), 1879–1888 (2009)
8. Jukes, T.H., Cantor, C.R.: Evolution of protein molecules. In: Munro, H. (ed.) Mammalian Protein Metabolism, vol. 3, pp. 21–132. Academic Press, New York (1969)
9. Lacey, M.R., Chang, J.T.: A signal-to-noise analysis of phylogeny estimation by neighbor-joining: insufficiency of polynomial length sequences. Math. Biosci. **199**(2), 188–215 (2006)
10. Le, T.: GitHub site for the INC and constrained - INC software (2019). https://github.com/steven-le-thien/INC
11. Le, T., Sy, A., Molloy, E., Zhang, Q., Rao, S., Warnow, T.: Using INC within divide-and-conquer phylogeny estimation - datasets (2019). https://databank.illinois.edu/datasets/IDB-8518809
12. Lefort, V., Desper, R., Gascuel, O.: FastME 2.0: a comprehensive, accurate, and fast distance-based phylogeny inference program. Mol. Biol. Evol. **32**(10), 2798–2800 (2015). https://doi.org/10.1093/molbev/msv150
13. Liu, K., et al.: SATé-II: very fast and accurate simultaneous estimation of multiple sequence alignments and phylogenetic trees. Syst. Biol. **61**(1), 90–106 (2012). https://doi.org/10.1093/sysbio/syr095

14. Maddison, W.: Gene trees in species trees. Syst. Biol. **46**(3), 523–536 (1997)
15. Mallo, D., De Oliveira Martins, L., Posada, D.: SimPhy: phylogenomic simulation of gene, locus, and species trees. Syst. Biol. **65**(2), 334–344 (2016). https://doi.org/10.1093/sysbio/syv082
16. Mirarab, S., Nguyen, N., Wang, L.S., Guo, S., Kim, J., Warnow, T.: PASTA: ultra-large multiple sequence alignment of nucleotide and amino acid sequences. J. Comput. Biol. **22**, 377–386 (2015)
17. Mirarab, S., Reaz, R., Bayzid, M.S., Zimmermann, T., Swenson, M.S., Warnow, T.: ASTRAL: genome-scale coalescent-based species tree estimation. Bioinformatics **30**(17), i541–i548 (2014)
18. Molloy, E.K., Warnow, T.: NJMerge: a generic technique for scaling phylogeny estimation methods and its application to species trees. In: Blanchette, M., Ouangraoua, A. (eds.) RECOMB-CG 2018. LNCS, vol. 11183, pp. 260–276. Springer, Cham (2018). https://doi.org/10.1007/978-3-030-00834-5_15
19. Molloy, E.K., Warnow, T.: Statistically consistent divide-and-conquer pipelines for phylogeny estimation using NJMerge. bioRxiv (2018). https://doi.org/10.1101/469130
20. Nelesen, S., Liu, K., Wang, L.S., Linder, C.R., Warnow, T.: DACTAL: divide-and-conquer trees (almost) without alignments. Bioinformatics **28**(12), i274–i282 (2012)
21. Price, M.N., Dehal, P.S., Arkin, A.P.: FastTree 2 - approximately maximum-likelihood trees for large alignments. PloS One **5**(3), 1–10 (2010)
22. Roch, S., Sly, A.: Phase transition in the sample complexity of likelihood-based phylogeny inference. Probab. Theory Relat. Fields **169**(1), 3–62 (2017)
23. Sayyari, E., Whitfield, J.B., Mirarab, S.: Fragmentary gene sequences negatively impact gene tree and species tree reconstruction. Mol. Biol. Evol. **34**(12), 3279–3291 (2017)
24. Stamatakis, A.: RAxML version 8: a tool for phylogenetic analysis and post-analysis of large phylogenies. Bioinformatics **30**, 1312–1313 (2014)
25. Swofford, D.L.: PAUP* (*Phylogenetic Analysis Using PAUP), Version 4a161 (2018). http://phylosolutions.com/paup-test/
26. Tavaré, S.: Some probabilistic and statistical problems in the analysis of DNA sequences. In: Lectures on Mathematics in the Life Sciences, vol. 17, pp. 57–86. American Mathematical Society (1986)
27. Warnow, T.: Divide-and-conquer tree estimation: opportunities and challenges. In: Warnow, T. (ed.) Bioinformatics and Phylogenetics. Springer (2019)
28. Warnow, T., Moret, B.M., St. John, K.: Absolute convergence: true trees from short sequences. In: Proceedings of the ACM-SIAM Symposium on Discrete Algorithms, pp. 186–195. Society for Industrial and Applied Mathematics (2001)
29. Zhang, Q., Rao, S., Warnow, T.: Constrained incremental tree building: new absolute fast converging phylogeny estimation methods with improved scalability and accuracy. Algorithms Mol. Biol. **14**(2), 2 (2019). https://rdcu.be/blBXm

Predicting Methylation from Sequence and Gene Expression Using Deep Learning with Attention

Alona Levy-Jurgenson[1,3](✉) [iD], Xavier Tekpli[2], Vessela N. Kristensen[2], and Zohar Yakhini[1,3]

[1] Computer Science Department,
Technion - Israel Institute of Technology,
3200003 Haifa, Israel
alonal@cs.technion.ac.il
[2] The Norwegian Radium Hospital, Oslo University Hospital,
Sognsvannsveien 20, 0372 Oslo, Norway
[3] School of Computer Science, Interdisciplinary Center,
Kanfei Nesharim St., 4610101 Hertsliya, Israel

Abstract. DNA methylation has been extensively linked to alterations in gene expression, playing a key role in the manifestation of multiple diseases, especially cancer. Hence, the sequence determinants of methylation and the relationship between methylation and expression are of great interest from a molecular biology perspective. Several models have been suggested to support the prediction of methylation status. These models, however, have two main limitations: (a) they are limited to specific CpG loci; and (b) they are not easily interpretable. We address these limitations using deep learning with attention. We produce a general model that predicts DNA methylation for a given sample in any CpG position based solely on the sample's gene expression profile and the sequence surrounding the CpG. Depending on gene-CpG proximity, our model attains a Spearman correlation of up to 0.84 for thousands of CpG sites on two separate test sets of CpG positions and subjects (cancer and healthy samples). Importantly, our approach, especially the use of attention, offers a novel framework with which to extract valuable insights from gene expression data when combined with sequence information. We demonstrate this by linking several motifs and genes to methylation activity, including Nodal and Hand1. The code and trained weights are available at: https://github.com/YakhiniGroup/Methylation.

Keywords: Genomics · Methylation · Deep learning · Attention

1 Introduction

DNA methylation is a chemical process that modifies DNA in living organisms and can significantly affect gene expression, mostly through the inhibition of

© Springer Nature Switzerland AG 2019
I. Holmes et al. (Eds.): AlCoB 2019, LNBI 11488, pp. 179–190, 2019.
https://doi.org/10.1007/978-3-030-18174-1_13

transcription. In humans, DNA methylation refers to the presence of a methyl group at a defined position of a cytosine and occurs mostly in CpG dinucleotides. It has been particularly shown to affect gene expression in gene promoter regions with relatively dense CpGs, known as CpG islands (CGI). When a large number of proximal CpGs are methylated, the transcription of nearby downstream genes may be inhibited.

DNA methylation plays a key role in disease development. Specifically, hypermethylation can lead to stable silencing of tumor suppressor genes [13]. This process has therefore been extensively observed and studied in the context of cancer [9,19]. While various forms of cancer are central to the discussion on DNA methylation, it has also been linked to other diseases and biological processes such as cardiovascular disease [8] and Alzheimer's Disease [12].

Currently, there are several methods available for measuring DNA methylation [15]. Some of these methods, however, require specialized protocols or a relatively large DNA sample size. Hence, depending on the required task, the costs could be significant and the data collection may be impractical. In addition, and perhaps more importantly, the link between gene expression and methylation is still an open-ended question and predictive analyses could provide insight into this relationship. In this work we set out to address both aspects of methylation.

This work includes the following contributions: (1) A practical tool that enables users to input any CpG position along with the sample's gene-expression profile, and obtain a prediction. (2) From a theoretical perspective, our results provide proof for a sharp link between sequence and expression and between local methylation events. We also unveil motifs and genes that the model identified as significant contributors to the prediction, specifically linking HAND1 and NODAL to methylation activity in the cohort analyzed. (3) We provide a novel model design and framework that support the combination of gene-expression data with genomic sequences to extract valuable molecular-level insights.

1.1   Related Work

Over the past decade, researchers have been investigating the use of machine learning for the prediction of methylation. In [1] and [5] the authors used classifiers such as Support Vector Machines (SVMs) and decision trees to determine the status of a given CpG using both structural and sequence-specific features. Similarly, [25] suggested a random forest classifier that uses features such as genomic position and neighbor methylation levels. The latter were noted as significant contributors but clearly require collecting partial methylation data. Others [17], have used a regression approach to predict continuous methylation levels across tissues, also using SVMs. While the use of regression is indeed more appropriate in the context of continuous methylation measurements, this approach requires extensive data collection from a source tissue. More recently, [23] used a deep learning model to predict whether a CpG was hypo-or hyper-methylated by using DNA patterns and topological features. The latter are human engineered features taken as input by the network model. Like previous methods, this model is limited to binary classification, and is specifically constrained to hypo-/hyper-methylation.

To conclude, the main limitations posed by previous models include: (1) The need to measure methylation in some (or all, in the case of learning between tissues) CpG sites. (2) Extensive use of human-engineered features. This not only incorporates human biases, but also prevents the model from unveiling novel representations. (3) Use of binary classifiers when in reality methylation levels are measured continuously, representing fractions of cells with any given status.

2 Approach

To address the limitations posed by the aforementioned methods, we suggest a deep learning model that does not require measuring methylation levels to obtain a prediction, provides continuous predictions rather than binary and uses attention mechanisms that enable us to uncover important representations learned by the model. Specifically, we predict methylation levels at a given CpG in a given sample based on three factors: (1) The sequence surrounding the CpG; (2) The sample's gene expression profile; and (3) The distance between the profiled genes and the CpG.

We use a generalized approach that can be applied to any set of CpGs. We avoid incorporating human-engineered features derived from the DNA sequence by using a Convolutional Neural Network (CNN) as a motif detector. We also do not manually select genes to include in the model, but rather incorporate the gene-expression profile by using three attention mechanisms that take the input context into consideration. That is - the attention mechanism is determined by the sequence around the CpG of interest, the distance between the CpG of interest to each of the genes and the gene-expression profile as a whole. We test our model on completely separate sets of CpG positions and subjects to ensure that our model can indeed generalize beyond the training data. Our method is fully available online[1].

2.1 Datasets

We used data from two cancer cohorts: (1) 782 breast cancer patients and (2) 498 prostate cancer patients. For each patient, we obtained two types of data: (a) gene expression data in RSEM normalized count for 17,997 genes (RNA-seq) and (b) methylation levels at 360,531 CpG sites (450K Illumina array).

In addition to the patient-specific data, we also use data specific to any CpG locus: (a) the ambient sequence - 399 base-pairs upstream of the CpG and 399 downstream, for a total of 800 base-pairs and (b) the genomic distance between each gene in the profile considered and the CpG of interest.

---

[1] https://github.com/YakhiniGroup/Methylation.

2.2 Constructing the Model

Our task is to predict the methylation level at a CpG site in a given sample using the samples's gene expression profile and the ambient sequence at the CpG site. To do so, we created a multi-modal neural network (Fig. 1) comprised of four sub-networks: one CNN, which acts as a motif detector for the surrounding sequence and three attention components which act as gene amplifiers, each based on the input provided. These sub-networks are then combined into a single fully-connected network to produce the final prediction.

We define a single training example to represent one subject (or sample) and one CpG. It contains the following components:

1. The subject's gene expression vector e, where each entry, e_i, represents the expression level of a gene g_i.
2. The sequence surrounding the CpG of interest, represented as a matrix S of one-hot vectors as shown in (Fig. 1).
3. A vector d, where d_i is computed based on the distance, in base-pairs, between g_i and the CpG of interest. Specifically, a gene residing within the first 2,000 base-pairs receives a value of 1, the next 2,000 a value of 0.5 and so on until the last bucket of 2,000 was given a value of 0.5^9. Beyond this point, and for genes residing on a different chromosome, this value was 0.

2.3 Attention Mechanism for Gene-Expression

To incorporate the gene-expression profile, we use attention mechanisms. An attention mechanism is essentially a vector of probabilities usually obtained by employing softmax on the final output layer of a neural network. This vector in turn is used as a filter for another vector, often via an element-wise product.

In our case, we created three attention vectors, each of which is derived from the output vector of a different neural network. We then multiplied each of them element-wise by the gene-expression profile vector e as seen in Fig. 1. Specifically, we created the following three neural networks to generate three attention vectors:

1. A second CNN that operates on S, with output layer a_{seq}. This is in addition to the CNN described earlier. The two allow the model to separate between motifs that are gene-specific (via attention) and general motifs (e.g. CG).
2. A fully-connected neural network based on the distance vector d with output layer a_{dist}.
3. A fully-connected neural network based on the gene-expression vector e with output layer a_{exp}.

These attention mechanisms enable the model to select which genes are important given any input context and provide a form of conditional importance to all measured expression levels. For example, the first attention vector might detect the presence of a transcription factor binding site (TFBS) proximal to the CpG via motifs learned by the CNN, indicating it may be related to methylation activity. The second attention mechanism might learn that a gene that is in close proximity to the given CpG has more predictive value than a gene residing on a separate chromosome. The third might detect that a certain combination of co-expressed genes is informative of the expression level of some other gene.

Each of the three attention vectors is multiplied element-wise by the gene-expression profile e to produce three vectors of the following form:

$$a_i \odot e \tag{1}$$

where $\odot$ is the element-wise product and a_i is one of the three attention vectors.

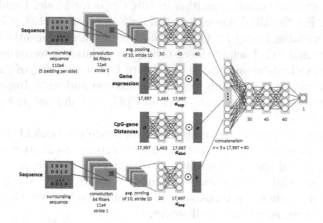

Fig. 1. The full multi-modal neural network, combining a CNN motif-detector (top) with three attention components applied on gene-expression levels (based on sequence, gene-expression and distance). The final layers of these sub-networks are combined via concatenation and fed into a final neural network. Input layers are in orange and the final output layer, representing the predicted level of methylation at the CpG of interest and in the sample of interest, is in green. (Color figure online)

2.4   The Combined Multi-modal Neural Network

We combine the output layers of the first CNN and the three attention mechanisms via concatenation and feed the concatenated representation into a final fully-connected neural network. More formally, denoting the output layer representing the surrounding sequence as s, we form the following input to the final fully-connected network (described in Fig. 1):

$$[s,\ a_{seq} \odot e,\ a_{dist} \odot e,\ a_{exp} \odot e] \tag{2}$$

2.5 Training

We create a unified model by training three different sub-models designed to address three close, but not identical, prediction tasks: Model 1 - focuses on CpGs with a gene that is in close proximity, specifically within a window of 2,000 base-pairs on either side; Model 2 - focuses on CpGs with a gene that is in medium-proximity of 10,000 base-pairs; and Model 3 - applicable to all CpGs regardless of gene proximity. To ensure that our model learns to generalize across different samples, we created a unified dataset that includes both the breast cancer and prostate cancer cohorts, as described in Sect. 2.1.

For each model we created a new dataset that matched the task's criteria. For the first two models we took all CpGs from the combined dataset that satisfied the relevant window criteria, resulting in 10K CpGs for Model 1 and 74K CpGs for Model 2. For the third model we randomly sampled 100K CpGs out of the total 360K available (using all CpGs would require a significantly larger gene-CpG distance matrix). Each of the resulting three datasets were randomly split into training, validation and test sets (65% of CpGs and 80% of subjects were train and the rest split randomly between validation and test). Importantly, the held-out test sets contain only subjects and CpGs that *did not participate in the training phase*.

The purpose of generating these three sub-models is twofold: (1) it enables us to provide more accurate predictions under certain proximity conditions and (2) having a model specialize solely on proximity data (as opposed to generic CpG data), especially in the case of Model 2 where many CpGs satisfy the 10,000 window condition, will enable it to learn more effectively those gene-CpG representations we sought out to discover.

Our models were trained using the Adam Optimizer [14]. We use the mean-absolute error (MAE) as the loss function. Training to convergence on the validation set took roughly 4 h on a Tesla K80 GPU.

3 Results

3.1 Predicting Methylation Levels

We evaluated each model on its respective held-out test sets, as described earlier. Model 1, for gene-proximal CpGs, achieves an MAE of 0.13 and 0.84 Spearman correlation ($-\log p > 100$). Model 2, for gene-neighboring CpGs, achieves an MAE of 0.16 and 0.75 Spearman correlation ($-\log p > 100$). Model 3, trained on general CpGs, attains 0.2 MAE and 0.65 Spearman correlation ($-\log p > 100$). These results demonstrate that the model better utilizes the attention mechanisms when provided with relevant data. Furthermore, these results also show that the models are capable of generalizing to new CpGs and subjects in the context of cancer (recall that the test sets do not include any CpGs or subjects from the training sets).

While our models were trained on two cancer cohorts, we also tested its performance on healthy tissue measurements (without any training on this data).

This would provide further indication of the ability of our model to generalize. For this purpose, we tested Model 1 on the corresponding healthy tissue samples available for the breast cancer cohort (n = 85). Out of the available 85 subjects, we kept only those that did not appear in the model's train or validation set. We did the same for CpGs, removing any that appeared in the train or validation sets. This resulted in 21 subjects out of the available 85 and 956 CpGs. Our model obtained 0.84 Spearman correlation ($-\log p > 100$) on this test set.

3.2   Comparison to Previous Work

We compare our model to two representative methods - a deep learning approach used for classification [23] and a feature-based approach used for regression [17]. The first method used a fully-connected neural network to classify hypo- and hyper- methylated CpGs. While the model obtains an accuracy of 80%–90% on this task, the approach is limited to extreme methylation levels. Furthermore, higher accuracy was obtained using methylation levels at neighboring CpGs, requiring pre-measuring methylation levels at those sites. In this case, it may be more practical to measure methylation levels directly at the CpGs of interest.

The second model used regression SVM to predict methylation levels at a target tissue based on a source tissue. Depending on which tissues were used for the task, they obtained R^2 of 0.73–0.99. The downside of this approach is the need to collect methylation data from a specific source tissue at specific CpGs.

Our model does not require collecting any prior measurements of methylation. It is also not limited to any specific range of prediction values. We demonstrated our model's ability to generalize by testing it on a completely separate and randomly selected set of CpGs and subjects, as well as on a separate test set obtained from healthy tissue. Most importantly, our model introduces the use of CNNs and attention to this task, both of which enable to extract insights from the learned representations using our proposed framework.

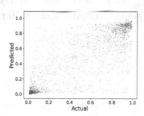

Fig. 2. Actual (x-axis) vs predicted methylation values for Model 1 (3,000 randomly selected instances from its held-out test set).

3.3   Gene-Expression Attention Learned by the Model

In this section we analyze the attention components, and demonstrate that they can provide valuable insights into the learned representations. For this purpose we used Model 2.

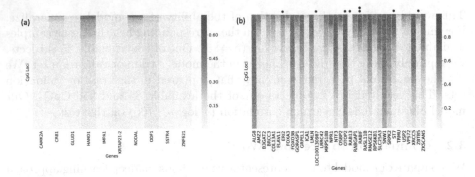

Fig. 3. Attention probabilities per gene (columns) and CpG (rows). Each gene's CpGs were sorted from high (blue) to low (yellow) attention scores. The first 100 CpGs per gene are shown. (a) Sequence-based attention - genes with at least one attention score >0.1. (b) Distance-based attention - genes with a median attention score >0.1. (Color figure online)

Sequence-Based Attention. In this section, we set out to analyze the genes that were attended to by the sequence-based attention mechanism. To do so, we first identified those genes that were relevant across multiple CpGs (i.e. attended to by multiple sequences). Specifically, we gathered all attention vector outcomes for each of the unique sequences in the test set, resulting in a single attention vector per CpG. We then sliced across CpGs and gathered all attention scores attributed to a gene (from all CpGs), resulting in one vector per gene. We sorted each of the gene's attention values in descending order so that the top of its list corresponds to those CpGs for which the gene received the highest attention score (importance). Finally, we removed all genes that did not have at least one attention score >0.1. The top 100 CpGs from each of the remaining sorted lists are the columns seen in Fig. 3(a). Note that each gene was sorted separately, hence the top CpGs in one gene's column may differ from those of another.

Fig. 4. The most significantly enriched motifs in sequences with high sequence-based attention scores for HAND1 (a) and NODAL (b). The HAND1 motif is near-identical to its known binding site motif GTCTGG.

The most notable gene is HAND1. To assess the possible relationship between HAND1 and DNA methylation in our dataset, we retrieved the sequences surrounding the top CpGs for HAND1 (recall that attention is determined based on the CpG's ambient sequence, in this case), and tested whether they are significantly enriched with the HAND1 binding site motif - GTCTGG [10] as compared to the lower ranked CpGs. Such a case would indicate that the CpGs proximal to this motif produce higher attention scores for HAND1, linking their methylation level to HAND1 binding. We performed this analysis using DRIMUST [16], a tool which analyzes motifs enriched at the top of ranked lists. We took the top 20 CpGs for HAND1, with an average attention value of 0.32, along with the bottom 20 CpGs, with an average attention value of 0.07, and inserted them in ranked order. The top significantly enriched motif - GTCTGA - was indeed nearly identical to the known HAND1 binding motif (p-value $< e^{-6}$), as can be seen in Fig. 4(a). Furthermore, only the top 18 CpGs contained this motif and the top 10 contained it more than once. None of the bottom 20 sequences contained this motif. To the best of our knowledge, the HAND1 gene has not been previously associated directly with the process of methylation at or near HAND1 binding sites, yet our findings show that this may be the case.

Another prominent gene, which was given a high attention score by multiple CpGs, is NODAL. NODAL plays a crucial part in the Nodal Signaling Pathway (NSP), in which it is responsible for instigating the transcription of multiple target genes. This is likely part of the reason why the model attributed NODAL with a high attention value for multiple CpGs. According to a recent experiment conducted on mouse embryos, elevated NODAL levels may be linked with increased DNA methylation [4]. Combining this finding with the outcome of our model indicates this may also be the case in humans, and specifically in cancer tissues. Performing motif analysis here, we discover a single significantly enriched motif - CGGCGGC (p-value $< e^{-10}$) as seen in Fig. 4(b). Here too, the motif appears only in the top 20 sequences, and multiple times in the vast majority of them (the top 8 alone contain 37 occurrences).

Distance-Based Attention. For the distance-based analysis, we obtained each gene's attention vector in a similar fashion to that described in the sequence-based section. This resulted in thousands of genes with at least one distance-based attention score larger than 0.1. Hence, for display purposes, we further refined the list of genes to include those for which the top 50 CpGs were larger than 0.1. This resulted in 36 genes for which the top 100 CpGs are shown in Fig. 3(b).

We hypothesized that the high attention scores may be explained by two main mechanisms: (1) in-cis effect: the CpGs reside within the 10,000 base-pair window of that gene and are therefore directly associated with it; or (2) in-trans effect: the gene is associated with multiple CpGs in different genomic locations. To examine the first hypothesis, we tested for a significant enrichment of nearby CpGs at the top of the attention list of each gene g in Fig. 3(b). This was accomplished using the minimum hypergeometric (mHG) test [6]. The analysis yielded 7 genes for which the high attention values were significantly enriched with CpGs residing within the window limit (p-value < 0.05). These genes are marked with (*) in Fig. 3(b). Specifically, RABIF had a p-value < 0.003 and is marked with (**).

To further explain the second hypothesis (in-trans), we will use FOXO3B as a simple example. Notice that FOXO3B is not marked with (*), implying that its high attention scores are not associated with nearby CpGs. Instead, the model might have learned that the expression level of FOXO3B is associated with methylation levels of a distant set of CpGs. Indeed, FOXO3B's top 20 CpGs (p-value < 0.0001) reside on 8 different chromosomes. This is consistent with the fact that it is a member of the forkhead family of DNA binding proteins [2].

3.4   Motifs Learned by the CNN

In this section we analyzed the representations learned by the first CNN (top of Fig. 1). Specifically, we took each learned CNN filter, and 0–1 scaled each row (representing a single nucleotide) to obtain the position probability matrix used for generating sequence logos. Figure 5 described six motifs detected by the CNN. One representation that stands out (Fig. 5(b)) indicates the importance of both the individual CpGs within the surrounding sequence, as well as the existence of multiple, consecutive combinations of Cs and Gs, most likely representing dense CpG occurrences - a hallmark of CpG islands.

Another prominent outcome is the CA-motif in Fig. 5(a). This motif has been shown to modulate alternative splicing of mRNA [11], which is also thought to be regulated by methylation [18]. CpA repeats (TpG repeats) are also hallmarks of past methylation activity due to conversion of CpG to TpG when deamination follows methylation [3].

We also identified the TATA motif (or TATA box), a core promoter element [24] seen in Fig. 5(c). Previous studies have shown that promoters residing in CpG-islands, and CG-dense regions in general, often lack this motif [24]. Hence, the model seems to distinguish between CpGs residing in CpG-dense regions and CpGs that are relatively isolated.

Also worth mentioning, is the SP1 motif (Fig. 5(e)) which is especially known for its consensus sequence: GGGCGG and its reverse complement [22].

4   Conclusions

DNA methylation is strongly related to disease development and is therefore the focus of much research. Models that provide methylation predictions could speed up future research and improve our understanding on how epigenetics may be involved in physiopathology. In this paper, we provided a general model that supplies such predictions based on the ambient sequence at the CpG of interest and the sample's gene expression profile. Our model is comprised of three sub-models that enable us to provide more accurate predictions under certain gene-CpG proximity conditions. We demonstrated the model's capability of generalizing across both CpGs and samples by testing on completely separate sets of CpGs and subjects, as well as on healthy tissue. Our model is highly interpretable, avoids incorporating prior knowledge by using CNNs and attention as feature extractors, provides continuous predictions and is not limited to any subset of CpGs, unlike previous models.

Most importantly, we demonstrate the power of using an attention mechanism on gene-expression data by analyzing its learned representations. Specifically, this enabled us to link HAND1 and NODAL to methylation activity.

Our attention-based model, along with its analysis, provide a novel framework for future research that seeks to combine gene-expression data with genomic sequences and extract valuable insights from both. This framework could also be extended beyond gene-expression data to include other genomic measurements.

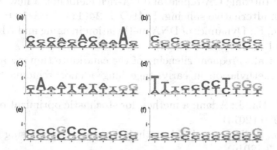

Fig. 5. Motifs identified by the CNN. (a) CA-repeat - a hallmark of methylation activity [20]; (b) CpG-dense regions; (c) TATA motif - a core promoter element [24]; (d), (e) SP1 motif, known for the consensus sequence: GGGCGG and its reverse complement [22] and for consecutive Cs or Ts [21]; (f) DNA repeats, also linked to methylation [7].

Acknowledgments. We would like to thank the Yakhini Group, and specifically Leon Anavy and Oz Solomon, for valuable discussions and suggestions. We also thank Anthony Mathelier and colleagues from the Kristensen Group for important comments.

References

1. Bhasin, M., et al.: Prediction of methylated CPGS in DNA sequences using a support vector machine. FEBS Lett. **579**(20), 4302–4308 (2005)
2. Chen, X., Ji, Z., Webber, A., Sharrocks, A.D.: Genome-wide binding studies reveal DNA binding specificity mechanisms and functional interplay amongst forkhead transcription factors. Nucl. Acids Res. **44**(4), 1566–1578 (2015)
3. Cooper, D.N., et al.: Methylation-mediated deamination of 5-methylcytosine appears to give rise to mutations causing human inherited disease in CpNpG trinucleotides, as well as in CpG dinucleotides. Hum. Genomics **4**(6), 406 (2010)
4. Dai, H.Q., et al.: TET-mediated DNA demethylation controls gastrulation by regulating Lefty-Nodal signalling. Nature **538**(7626), 528 (2016)
5. Das, R., et al.: Computational prediction of methylation status in human genomic sequences. Proc. Nat. Acad. Sci. **103**(28), 10713–10716 (2006)
6. Eden, E., et al.: Discovering motifs in ranked lists of DNA sequences. PLoS Comput. Biol. **3**(3), e39 (2007)
7. Ehrlich, M.: DNA methylation in cancer: too much, but also too little. Oncogene **21**(35), 5400 (2002)
8. Fiorito, G., et al.: Oxidative stress and inflammation mediate the effect of air pollution on cardio - and cerebrovascular disease: a prospective study in nonsmokers. Environ. Mol. Mutagen. **59**(3), 234–246 (2018)

9. Grasso, C.S., et al.: Genetic mechanisms of immune evasion in colorectal cancer. Cancer Discov. **8**, 730–749 (2018)

10. Hollenberg, S.M., et al.: Identification of a new family of tissue-specific basic helix-loop-helix proteins with a two-hybrid system. Mol. Cell. Biol. **15**(7), 3813–3822 (1995)

11. Hui, J., et al.: Intronic CA-repeat and CA-rich elements: a new class of regulators of mammalian alternative splicing. EMBO J. **24**(11), 1988–1998 (2005)

12. Irier, H.A., Jin, P.: Dynamics of DNA methylation in aging and Alzheimer's disease. DNA Cell Biol. **31**(S1), S-42 (2012)

13. Kajiura, K., et al.: Frequent silencing of the candidate tumor suppressor TRIM58 by promoter methylation in early-stage lung adenocarcinoma. Oncotarget **8**(2), 2890 (2017)

14. Kingma, D.P., Ba, J.: Adam: a method for stochastic optimization. arXiv preprint arXiv:1412.6980 (2014)

15. Kurdyukov, S., Bullock, M.: DNA methylation analysis: choosing the right method. Biology **5**(1), 3 (2016)

16. Leibovich, L., et al.: Drimust: a web server for discovering rank imbalanced motifs using suffix trees. Nucl. Acids Res. **41**(W1), W174–W179 (2013)

17. Ma, B., et al.: Predicting DNA methylation level across human tissues. Nucl. Acids Res. **42**(6), 3515–3528 (2014)

18. Maor, G.L., et al.: The alternative role of DNA methylation in splicing regulation. Trends Genet. **31**(5), 274–280 (2015)

19. Nejman, D., et al.: Molecular rules governing de novo methylation in cancer. Cancer Res. **74**(5), 1475–1483 (2014)

20. Nichol, K., Pearson, C.E.: CpG methylation modifies the genetic stability of cloned repeat sequences. Genome Res. **12**(8), 1246–1256 (2002)

21. Plumitallo, S., et al.: Functional analysis of a novel eng variant in a patient with hereditary hemorrhagic telangiectasia (HHT) identifies a new Sp1 binding-site. Gene **647**, 85–92 (2018)

22. Raiber, E.A., et al.: A non-canonical DNA structure is a binding motif for the transcription factor Sp1 in vitro. Nucl. Acids Res. **40**(4), 1499–1508 (2011)

23. Wang, Y., et al.: Predicting DNA methylation state of CpG dinucleotide using genome topological features and deep networks. Sci. Rep. **6**, 19598 (2016)

24. Yang, C., et al.: Prevalence of the initiator over the tata box in human and yeast genes and identification of DNA motifs enriched in human tata-less core promoters. Gene **389**(1), 52–65 (2007)

25. Zhang, W., et al.: Predicting genome-wide DNA methylation using methylation marks, genomic position, and DNA regulatory elements. Genome Biol. **16**(1), 14 (2015)

A Mathematical Model for Enhancer Activation Kinetics During Cell Differentiation

Kari Nousiainen(✉) ⓘ, Jukka Intosalmi ⓘ, and Harri Lähdesmäki

Department of Computer Science, Aalto University School of Science,
00076 Aalto, Finland
kjnousia@gmail.com

Abstract. Cell differentiation and development are for a great part steered by cell type specific enhancers. Transcription factor (TF) binding to an enhancer together with DNA looping result in transcription initiation. In addition to binding motifs for TFs, enhancer regions typically contain specific histone modifications. This information has been used to detect enhancer regions and classify them into different subgroups. However, it is poorly understood how TF binding and histone modifications are causally connected and what kind of molecular dynamics steer the activation process.

Contrary to previous studies, we do not treat the activation events as static epigenetic marks but consider the enhancer activation as a dynamic process. We develop a mathematical model to describe the dynamic mechanisms between TF binding and histone modifications known to characterize an active enhancer. We estimate model parameters from time-course data and infer the causal relationships between TF binding and different histone modifications. We benchmark the performance of this framework using simulated data and survey the ability of our method to identify the correct model structures for a variety of system dynamics, noise levels and the number of measurement time points.

Keywords: Dynamic modeling · Enhancer activation ·
Cell differentiation

1 Introduction

Cell differentiation is steered by highly complex molecular machinery which controls the execution of cell type specific transcriptional programs. Transcriptional programs are typically initiated by external signaling molecules which bind receptors on the cell surface and activate downstream signaling pathways. Signaling cascades in turn activate lineage determining transcription factors that bind selected regions in the genome and control the regulatory functions of these regions, for instance, by modifying the structure of chromatin. An important class of genomic regions bound by lineage specifying transcriptions factors

© Springer Nature Switzerland AG 2019
I. Holmes et al. (Eds.): AlCoB 2019, LNBI 11488, pp. 191–202, 2019.
https://doi.org/10.1007/978-3-030-18174-1_14

are so-called enhancer regions which facilitate DNA looping, and as a consequence, enable interaction between enhancer and promoter sequences leading to the initiation of transcription [7,23]. Systematic activation of these selected regulatory regions is an important part of lineage determine supervision, but the detailed molecular kinetics behind these epigenetic mechanisms are poorly understood [15]. Modern high-throughput measurement techniques, such as chromatin immunoprecipitation sequencing, provide practical means to observe the epigenetic states in different cell types. Such methods have vigorously been used to determine global chromatin markings across different cell types. Further, new insights on the molecular mechanisms can be gained by analyzing these data using computational approaches.

Computational methods that have been used to study epigenetic mechanisms include, for instance, Bayesian networks, sparse partial correlation networks and maximum entropy framework [12,13,17,25]. Yu et al. [24] have also combined gene expression data analysis with histone modification networks and theoretical investigations of histone modification networks have also been proposed in [9]. Even though these existing approaches provide invaluable information about the structure of epigenetic signaling networks, the approaches are limited in the sense that they provide only a static view on the network structure and do not account for dynamic causative relationships between epigenetic modifications. In other words, these approaches provide information about statistical dependencies between measured quantities but are incapable of capturing mechanistic features of the underlying system. Dynamic view on epigenetic signaling is especially important when considering cell differentiation processes which depend on enhancer activity.

During cell differentiation processes, the state of an enhancer typically changes transiently from an inactive state to an active state. The activation process is steered by enzymatic signals (readers and writers of histone modifications) accompanied with appropriate transcription factor activities [2]. Consequently, in order to learn dynamic behavior leading to active enhancer state, it is necessary to quantify the dynamic and causative relationships between the key components of this complex molecular system.

In this study, we take the first steps towards mechanistic analysis of epigenetic signaling events that lead to the enhancer activation. We construct a mechanistic ordinary differential equation model to describe central histone modification and transcription factor dynamics leading to active enhancer state. Our model is designed to capture the dynamics of histone H3 lysine 4 monomethylation (H3K4me1), histone H3 lysine 27 acetylation (H3K27ac), and an activating TF. Highly enriched levels of histone modifications H3K4me1 [10] and H3K27ac [3] are known to characterize the active enhancer state e.g. in Th2 cells and known TFs drive the differentiation into Th2 lineage [16]. To carry out the analysis in a data driven manner, we embed the causal model into a statistical framework which makes it possible to infer the model structure as well as parameters from experimental data. Our result show that experimental data from as few as five time point is sufficient to distinguish cascade of enhancer activation events.

2   Methods

2.1   Mathematical Model

Lineage determining transcription factors, such as STAT family, play a crucial role in the enhancer activation as well as in the differentiation of Th cells in general [22]. One such scenario leading to enhancer activation is illustrated in Fig. 1. Initially, there are no TFs bound to the enhancer site and the enhancer associated histone modifications are absent (Fig. 1, initial state 1). As the first step, enhancer activation is initiated by TF that binds the enhancer (Fig. 1, state 2) and, in the considered scenario, TF binding is first followed by H3K4me1 (Fig. 1, state 3) and then finally by the third activation event H3K27ac (Fig. 1, state 4). This chain of three causative activation steps leads to active enhancer state and if the underlying assumptions of the causal relationships between the key components are correct, we can build a dynamic model for the activation process by using ODEs that describe the key components.

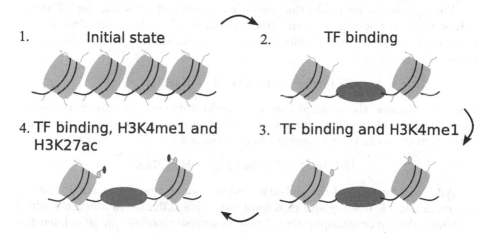

Fig. 1. A possible pathway model of chromatin changes during enhancer activation that are reflected by the abundance data.

Because TF activation is driven by T cell activation as well as inducing cytokine signals, we can simply assume that there is a persistent input signal affecting TF. Thus, TF dynamics can be described by the ordinary differential equation

$$\frac{d\,[\text{TF}]}{dt} = \alpha_{\text{TF}} - \delta_{\text{TF}}[\text{TF}], \tag{1}$$

where α_{TF} and δ_{TF} are unknown association and dissociation rate constants of TF and [TF] represent the TF abundance at the enhancer site. Further, if we assume that enzymatic signals that cause the methylation of the histone tail

H3K4me1 result from TF binding, we can model H3K4me1 enrichment at the enhancer cite by means of the equation

$$\frac{d[\text{H3K4me1}]}{dt} = \alpha_{\text{me}}[\text{TF}] - \delta_{\text{me}}[\text{H3K4me1}], \tag{2}$$

where α_{me} and δ_{me} are unknown methylation and demethylation rate constants and [H3K4me1] represent the H3K4me1 abundance at the enhancer site. Similarly, if H3K27ac is driven by H3K4me1 driven enzymatic signals, its enrichment [H3K27ac] at the site can be modeled using the differential equation

$$\frac{d[\text{H3K27ac}]}{dt} = \alpha_{\text{ac}}[\text{H4K3me1}] - \delta_{\text{ac}}[\text{H3K27ac}], \tag{3}$$

where α_{ac} and δ_{ac} are unknown acetylation and deacetylation rate constants. If the causative relationships between the key component TF, H3K4me1, and H3K27ac are correct, the resulting ODE system can be used to approximate the complex molecular kinetics consisting, for example, of TF binding events, enzymatic signals, methylation, acetylation etc.

The approximative model that we derive above indicates that the TF signal drives H3K4me1 which, in turn, drives H3K27ac. In other words, we have a cascade of causative events leading to the active enhancer state and this can be denoted by writing

$$\text{TF binding} \rightarrow \text{H3K4me1} \rightarrow \text{H3K27ac}.$$

However, because the detailed kinetics remain unknown, we cannot be sure if the causative relationships are correct. In other words, the true order of the activation events in the cascade can be, for instance,

$$\text{H3K4me1} \rightarrow \text{TF binding} \rightarrow \text{H3K27ac}.$$

Additionally, some of the activating events can also be synergistic in nature. For example, TF binding and H3K4me1 can drive H3K27ac in a manner which is either additive or multiplicative. These synergistic models can be defined formally using the following equations

$$\frac{d[\text{H3K27ac}]}{dt} = \alpha_{\text{ac1}}[\text{TF}] + \alpha_{\text{ac2}}[\text{H3K4me1}] - \delta_{\text{ac}}[\text{H3K27ac}], \tag{4}$$

and

$$\frac{d[\text{H3K27ac}]}{dt} = \alpha_{\text{ac}}[\text{TF}][\text{H3K4me1}] - \delta_{\text{ac}}[\text{H3K27ac}]. \tag{5}$$

All different scenarios that can be derived by altering the order or the type of the activation steps can be modeled using the rate equations given in Eqs. 1–5. Altogether there are 13 effectively different alternative models (see Table 1). Importantly, all these alternative scenarios result in different dynamics of the model output. Further, the dynamics of the model output can be directly linked to observed time-course data and, by means of quantitative statistical methods, it is possible to infer the most likely causative relationships between the key components leading to enhancer activation.

Table 1. A $\perp\!\!\!\perp$B denotes that events A and B are independent; A $\rightarrow$ B denotes that A regulates B; (A $\vee$ B) denotes A or B; (A $\wedge$ B) denotes A and B.

Index	Type of interaction	Order of events
0	Independent	H3K4me1 $\perp\!\!\!\perp$H3K27ac $\perp\!\!\!\perp$TF binding
1	Cascade	H3K27ac $\rightarrow$ H3K4me1 $\rightarrow$ TF binding
2	Cascade	H3K27ac $\rightarrow$ TF binding $\rightarrow$ H3K4me1
3	Cascade	H3K4me1 $\rightarrow$ H3K27ac $\rightarrow$ TF binding
4	Cascade	H3K4me1 $\rightarrow$ TF binding $\rightarrow$ H3K27ac
5	Cascade	TF binding $\rightarrow$ H3K27ac $\rightarrow$ H3K4me1
6	Cascade	TF binding $\rightarrow$ H3K4me1 $\rightarrow$ H3K27ac
7	Multiplicative synergy	(H3K4me1 $\wedge$ H3K27ac) $\rightarrow$ TF binding
8	Multiplicative synergy	(H3K4me1 $\wedge$ TF binding) $\rightarrow$ H3K27ac
9	Multiplicative synergy	(H3K27ac $\wedge$ TF binding) $\rightarrow$ H3K4me1
10	Additive	(H3K4me1 $\vee$ H3K27ac) $\rightarrow$ TF binding
11	Additive	(H3K4me1 $\vee$ TF binding) $\rightarrow$ H3K27ac
12	Additive	(H3K27ac $\vee$ TF binding) $\rightarrow$ H3K4me1

2.2 Statistical Framework

We combine the dynamic ODE models with time-course data by means of statistical modeling. More specifically, we set up a statistical framework for the ODE models by using Bayesian methodology as outlined in [8] and carry out posterior inference for the parameters and for the most likely causative relationships between the key components steering enhancer activation [6] (alternative models are listed in Table 1). In the following, we describe the details of our statistical framework.

Let us denote the output of the model M_k by $\phi_{M_k}(\theta_k, t) \in \mathbb{R}^N$ where N is the number of components in the model, θ_k is the vector of parameters of the model and t is the time point. Further, θ_{kl} is the l'th element of θ_k. Also, let $\mathcal{D} = (y_{11}, \ldots, y_{NT})$ be the experimental data which consists of measurements y_{ij} of the components $i = 1, \ldots, N$ at the time points $t = 1, \ldots, T$. Accordingly, $\phi_{ik}(\theta_k, t_j)$ is the is i'th element at j'th time point of the model output. By assuming normal errors, we define likelihood as

$$p(\mathcal{D} \mid M_k, \theta_k) = \prod_{i=1}^{N} \prod_{j=1}^{T} \mathcal{N}(y_{ij} \mid \phi_{ik}(\theta_k, t_j), \sigma_k^2)$$

where $\mathcal{N}$ is the normal probability density function with mean $\phi_{ik}(\theta_k, t_j)$ and variance σ_k^2. Posterior distribution of the model M_k is

$$p(M_k \mid \mathcal{D}) \propto p(\mathcal{D} \mid M_k)\pi(M_k)$$

or

$$\log(p(M_k \mid \mathcal{D})) = \log(p(\mathcal{D} \mid M_k)) + \log(\pi(M_k)) + C$$

where $\pi(M_k)$ is the prior distribution for the model M_k and C is constant. The marginal likelihoods

$$p(\mathcal{D} \mid M_k) = \int p(\mathcal{D} \mid M_k, \boldsymbol{\theta}_k) \pi(\boldsymbol{\theta}_k \mid M_k) d\boldsymbol{\theta}_k$$

is used to compare models with respect to each other. It can be approximated in many ways [4]. In this study, we apply Laplace approximations (as described e.g. [1]). Assuming uniform prior distribution for the models, i.e. prior probabilities $\pi(M_k)$ are equal, for all $k = 0, \ldots, 12$, we obtain Bayesian information criterion [20] defined by

$$BIC = \log(p(D \mid \hat{\theta}_k, M_k)) - \frac{1}{2} k_0 \log(n),$$

where $p(D \mid \hat{\theta}_k, M_k)$ is the maximum likelihood for model M_k, k_0 is the number of parameters and n is the number of observations.

2.3   Computational Implementation

In this work, we applied tools and methods reported being successful in dynamical modeling in systems biology. The ODE models and the model selection were implemented in Matlab (The MathWorks Inc., Natick, MA, USA) by using PESTO toolbox [21] for parameter optimization and AMICI toolbox [5] for solving the ODE systems numerically. Maximum likelihood estimates for parameters were obtained by employing the sensitivity equations in combination with a multi-start strategy based on latin hypercube sampling as suggested in [18,19]. We optimized the parameters by maximizing the log-likelihood function from 100 starting points with interior-point algorithm in Matlab's fmincon function.

3   Results

3.1   Evaluation of Model Identifiability and Discrimination

In practice, time-course measurements for histone modifications and TF binding can be carried out only at a few time points. Being aware of the limited size of real data sets, we use a small number of time-points also in our experiments with simulated data. In these experiments, we consider three different scenarios for measurements time-point selection, in the first one the samples are collected at three time points (0, 4 and 72 h), in the second one the samples are collected at four time points (0, 4, 8, and 72 h), and in the third samples are collected at five time points (0, 4, 8, 12 and 72 h). For each scenario we considered eight levels of measurement noise. The ladder was used to estimate the upper limit of heterogeneity in enhancer activation signals supporting this approach. In addition, we introduced additional variability between the data sets by drawing the model parameters from normal distributions with fixed means shown in Table 2 and five percent coefficient of variation. This leads to heterogeneous data containing dynamics of varying rates. In total, we created independently 4800 different data sets.

We evaluated the model selection in two settings. First one was designed to be as flexible as possible. Model parameters, initial values and measurement noise variance all inferred from simulated data. The rate parameters were constrained to the range $[10^{-3}, 100]$, initial values at range $[0.05, 0, 5]$ and standard deviations $[0.05, 3]$. Model selection is shown in Fig. 2. Not surprisingly model selection is relatively unreliable with three times points. Yet, with four time points cascade model starts to become recognizable, when measurement noise is reasonable small (standard deviation less than, say, 1), Using five measurement times enables reliable model selection for cascade and additive models. However, it is difficult for this framework to identify the correct model from very limited amounts of data when synergistic model is used to generate the data. Moreover, the framework with this amount of flexibility, seems to favour models with some dependencies between the enhancer activation signals even when data is created from models consisting independent variables.

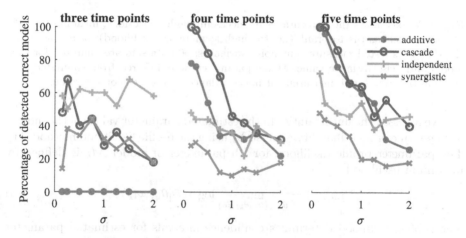

Fig. 2. Y axis shows the percentages of successful model selections, meaning that the correct model has the top rank (i.e. the highest marginal likelihood) for a given data. X axis shows tested measurement noise variances. 50 data sets are analyzed for each model and noise variance value. Model parameters, initial values and measurement noise variance were all inferred from simulated data.

The other setting is more rigid. In this scenario, only rate parameters are inferred from the data while initial values and values for measurements noise are fixed to the correct values. In experimental work, it may be difficult to fix these parameters to exactly correct values, but it increases the number of data points relative to the number of inferred parameters making model ranking theoretically more sound. Model selection results are shown in Fig. 3. The pattern is similar to the previous case where data from three or four time points were not sufficient to trace the correct model whereas model ranking starts to become more accurate when data is available from five time points.

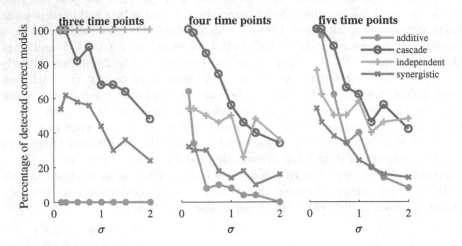

Fig. 3. Y axis shows the percentages of successful model selections, meaning that the correct model has the top rank (i.e. the highest marginal likelihood) for a given data. X axis shows tested measurement noise variances. 50 data sets are analyzed for each model and noise variance value. Model parameters were inferred from simulated data while initial values and measurement noise variance were set to correct value.

We assessed the uncertainties in the estimated parameter values of the models with parameter identifiability analysis by using the profile likelihood method [11]. The parameter profile likelihood for lth parameter of model $\mathcal{M}_k$ is defined by maximum likelihood

$$\text{PL}_{\theta_{kl}}(c) = \max_{\theta_k \in \{\theta_k | \theta_{kl} = c\}} \log p\left(\mathcal{D} | \theta_k, \mathcal{M}_k\right). \tag{6}$$

The profile likelihood determines confidence intervals for estimated parameter values given a fixed confidence level [11,14]. Profile qualifies whether the parameter is identifiable or not. There are three basic categories for parameter identifiability. Clean, smooth profile with obvious maximum which furthermore has reasonably constrained confidence interval is clearly identifiable. A flat profile leads to an infinite confidence interval indicating structural non-identifiability whereas a confidence interval constrained only at one end is practically non-identifiable.

As a likelihood based method identifiability analysis clearly depends on both data and measurement noise. We applied profile likelihoods to simulated data containing five time points where measurement noise has standard deviation 0.15. With this kind of data, parameters were generally identifiable for all models except for synergistic model family. Figure 4 illustrates this case. Even then most of the parameters were identifiable. Surprisingly, parameters of a synergistic model calibrated to data derived from cascade model were all identifiable.

Moreover, the same model calibration applied to additive model resulted in one practically non-identifiable parameter while others were indentifiable. Together, the results indicate that the inference framework performs well and model ranking is promisingly powerful.

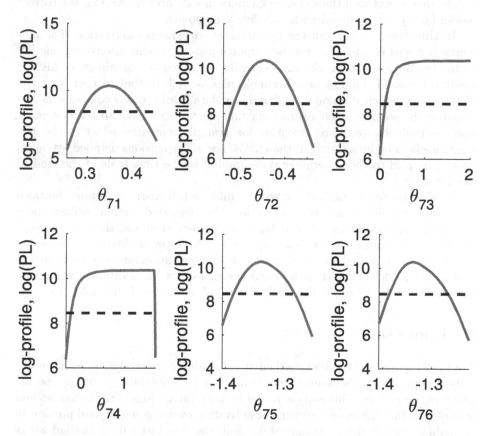

Fig. 4. The red curves represent the profile likelihoods of the kinetic parameters θ_7 represented in base 10 logarithmic scale. The dashed lines show the 95% confidence interval thresholds. The synergistic multiplicative model (model 7, Table 1.) was calibrated to data generated with the same model. Basal activation rate θ_{71} and deactivation rate θ_{72} define independent dynamics for the first activation signal x. Similarly, θ_{73} and θ_{74} control the dynamics of the second signal y. Signals x and y together drive the dynamics of z multiplicatively using parameter θ_{75} whereas θ_{76} is the deactivation rate of z. While other parameters are indentifiable, θ_{73} and θ_{74} are practically unidentifiable. (Color figure online)

4   Discussion

We propose a new computational framework which is based on network inference, representative systems of ODEs, parameter estimation and model ranking to infer and predict enhancer activation dynamics mechanistically. We verify that a feasible amount of data is able to distinguish different models by creating synthetic data that describes different kinds of dynamics and sample data points to simulate data from wet-lab experiments. When the data is sampled in time points that reflect well the dynamic changes in enhancer activation, the correct model family can be found with only five time points.

In this work, we concern the dynamics of an enhancer activation. The networks involved consist only the best known histone modifications and one TF factor binding. The networks can be easily expanded to contain other histone modifications, TF binding or other molecules as well. In that case, it is important to have sufficient amount of abundance data from the new components and consider the possible mechanisms the components can impact to the system. It may be that new terms representing for example repressive effect to the systems needs to be introduced to the ODEs. For all expansions detailed attention should be paid to the experimental design and to the time scale of the studied phenomenon.

Unlike previous methods used to infer mechanistic relations between molecules engaged in enhancer activation, the suggested method utilizes more effectively time evolution of abundance data. Instead of snapshots or series of snapshots the occasional relations captured by for example (dynamic) Bayesian networks, the proposed approach combines all information into a single complete dynamical model of relations between the molecules. In addition, mechanistic modeling enables us to predict the (relative) abundances of the molecules.

5   Conclusion

We have represented the first mathematical model to assess dynamic and mechanistic dependencies between key molecular components during cell type specific enhancer activation. This approach enables both predictions of the dynamics and model structure inference between the activation events going beyond previously available methods detecting only static features. The introduced method works well with data from a few time points and hence is applicable in both designing time course experiments and analyzing experimental data studies.

Acknowledgements. We acknowledge the computational resources provided by Aalto Science-IT project.

Funding. This work has been supported by the Academy of Finland, project 275537 and Chan Zuckerberg Initiative.

Appendix

Table 2. The simulated models and the means of the parameters used in data simulations. One representative from each model family were selected for generating data. Index $k \in \{0, 1, 7, 11\}$ specifies the model structure as described in Table 1. A sampled parameter vector θ_k consists of kinetic parameters θ_{kl}, initial values for three ordered enhancer activation signals denoted by x_0, y_0 and z_0 and the simulation specific measurement noise σ_s which was 0.15, 0.25, 0.5, 0.75, 1, 1.25 1.5 or 2.0. Independent, cascade and synergistic models have six kinetic parameters. Consecutive odd and even elements are the activation rates and the corresponding deactivation rates of the enhancer activation signal, respectively. Additive models have seven kinetic parameters. First four of them are the basal activation and the deactivation rates of enhancer activity signal x and y mediating dynamics independent from other variables while θ_{k5} and θ_{k6} are the activation rates of z activation caused by x and y and θ_{k7} is the deactivation rate of z.

Model	k	θ_{k1}	θ_{k2}	θ_{k3}	θ_{k4}	θ_{k5}	θ_{k6}	θ_{k7}	x_0	y_0	z_0	σ
Independent	0	3.1	.3	3.1	.3	3.1	.3	–	0.03	0.02	0.04	σ_s
Cascade	1	3.1	.3	.3	.25	.9	1	–	0.03	0.02	0.04	σ_s
Synergistic	7	2.4	.4	2	1	0.05	.05	–	0.03	0.02	0.04	σ_s
Additive	11	3.1	.3	.2	0.01	.9	0.7	1	0.03	0.02	0.04	σ_s

References

1. Bishop, C.M.: Pattern Recognition and Machine Learning. Information Science and Statistics. Springer, Berlin (2006)
2. Calo, E., Wysocka, J.: Modification of enhancer chromatin: what, how, and why? Mol. Cell **49**(5), 825–837 (2013)
3. Creyghton, M.P., et al.: Histone H3K27AC separates active from poised enhancers and predicts developmental state. Proc. Nat. Acad. Sci. **107**(50), 21931–21936 (2010)
4. Friel, N., Wyse, J.: Estimating the evidence-a review. Stat. Neerl. **66**(3), 288–308 (2012)
5. Fröhlich, F., Kaltenbacher, B., Theis, F.J., Hasenauer, J.: Scalable parameter estimation for genome-scale biochemical reaction networks. PLoS Comput. Biol. **13**(1), e1005331 (2017)
6. Gelman, A., et al.: Bayesian Data Analysis, vol. 2, 2nd edn. Chapman & Hall/CRC, Boca Raton (2004)
7. Ghavi-Helm, Y., et al.: Enhancer loops appear stable during development and are associated with paused polymerase. Nature **512**(7512), 96–100 (2014)
8. Girolami, M.: Bayesian inference for differential equations. Theor. Comput. Sci. **408**(1), 4–16 (2008)
9. Hayashi, Y., et al.: Theoretical framework for the histone modification network: modifications in the unstructured histone tails form a robust scale-free network. Genes Cells **14**(7), 789–806 (2009)

10. Heintzman, N.D., et al.: Distinct and predictive chromatin signatures of transcriptional promoters and enhancers in the human genome. Nat. Genet. **39**(3), 311–318 (2007)
11. Kreutz, C., Raue, A., Kaschek, D., Timmer, J.: Profile likelihood in systems biology. FEBS J. **280**(11), 2564–2571 (2013)
12. Lasserre, J., et al.: Finding associations among histone modifications using sparse partial correlation networks. PLoS Comput. Biol. **9**(9), e1003168 (2013)
13. Le, N.T., Ho, T.B.: Reconstruction of histone modification network from next-generation sequencing data. In: 2011 IEEE 11th International Conference on Bioinformatics and Bioengineering (BIBE), pp. 181–188. IEEE (2011)
14. Meeker, W.Q., Escobar, L.A.: Teaching about approximate confidence regions based on maximum likelihood estimation. Am. Stat. **49**(1), 48–53 (1995)
15. Natoli, G.: Maintaining cell identity through global control of genomic organization. Immunity **33**(1), 12–24 (2010)
16. Oki, S., Otsuki, N., Kohsaka, T., Azuma, M.: Stat6 activation and Th2 cell proliferation driven by CD28 signals. Eur. J. Immunol. **30**(5), 1416–1424 (2000)
17. Perner, J., et al.: Inference of interactions between chromatin modifiers and histone modifications: from chip-seq data to chromatin-signaling. Nucleic Acids Res. **42**(22), 13689–13695 (2014)
18. Raue, A., et al.: Lessons learned from quantitative dynamical modeling in systems biology. PloS One **8**(9), e74335 (2013)
19. Raue, A., et al.: Data2Dynamics: a modeling environment tailored to parameter estimation in dynamical systems. Bioinformatics **31**(21), 3558–3560 (2015)
20. Schwarz, G.: Estimating the dimension of a model. Ann. Stat. **6**(2), 461–464 (1978)
21. Stapor, P., et al.: PESTO: parameter estimation toolbox. Bioinformatics **34**(4), 705–707 (2018). https://doi.org/10.1093/bioinformatics/btx676
22. Vahedi, G., et al.: Stats shape the active enhancer landscape of T cell populations. Cell **151**(5), 981–993 (2012)
23. Voss, T.C., Hager, G.L.: Dynamic regulation of transcriptional states by chromatin and transcription factors. Nat. Rev. Genet. **15**(2), 69–81 (2014)
24. Yu, H., et al.: Inferring causal relationships among different histone modifications and gene expression. Genome Res. **18**(8), 1314–1324 (2008)
25. Zhou, J., Troyanskaya, O.G.: Global quantitative modeling of chromatin factor interactions. PLoS Comput. Biol. **10**, e1003525 (2014)

Transcript Abundance Estimation
and the Laminar Packing Problem

Atif Rahman[1,2](✉)(iD) and Lior Pachter[2,3,4](iD)

[1] Department of Computer Science and Engineering,
Bangladesh University of Engineering and Technology, Dhaka, Bangladesh
atif@cse.buet.ac.bd
[2] Department of Electrical Engineering and Computer Sciences,
University of California, Berkeley, Berkeley, CA, USA
[3] Departments of Mathematics and Molecular and Cell Biology,
University of California, Berkeley, Berkeley, CA, USA
[4] Departments of Biology and Computing and Mathematical Sciences,
California Institute of Technology, Pasadena, CA, USA
lpachter@caltech.edu

Abstract. The expectation-maximization (EM) algorithm, or a streaming version of it, is widely used to resolve ambiguity during transcript abundance estimation from RNA-seq reads. The streaming algorithm is fast and memory efficient but its accuracy can depend on the order of the reads, which can be stabilized if a tree can be constructed to capture the ambiguity structure. Motivated by this, the *laminar packing problem* is introduced, which is proved to be NP-hard. Hardness of approximation and approximability results are also provided. Finally, an integer linear programming (ILP) formulation and a greedy approach are applied to real data from the human transcriptome to demonstrate that large instances can be solved in practice.

Keywords: Laminar packing · Set packing · Independent set · Transcript abundance estimation · RNA-seq

1 Introduction

Transcript abundance estimation is the process of estimating relative quantities of transcripts in a cell or a set of cells from RNA-seq data. RNA-seq reads are first assigned to transcripts using a short read aligner as in [9,13], in an alignment free manner using k-mers [11], or through a process called pseudo-alignment [1,10]. A key issue is inferring the origin of ambiguously mapping sequenced reads. This is usually addressed using the expectation-maximization (EM) algorithm [1,9,11,13] or a streaming (online) version of it [10,12].

The online algorithm has the advantage that it is linear in the number of reads, is memory efficient, and can estimate abundance in real time without storing the reads. However, the accuracy of the estimates it provides can depend

© Springer Nature Switzerland AG 2019
I. Holmes et al. (Eds.): AlCoB 2019, LNBI 11488, pp. 203–211, 2019.
https://doi.org/10.1007/978-3-030-18174-1_15

on the order of the reads. For example, if numerous reads align to a transcript early during the online algorithm, later reads which are ambiguous with other transcripts may be preferentially assigned to the initial transcript just because it received reads first.

Motivated by this problem, we considered a stabilization procedure in the form of identifying a tree structure that can represent the ambiguity structure of transcripts (see Sect. 2 for details). In this tree, internal nodes correspond to sets of transcripts and ambiguous reads can be assigned to them so that final assignment to individual transcripts can be delayed. However, the ambiguity structure of transcripts may not be in the form of a tree. To address this issue we formulate the *laminar packing problem* which is the problem of constructing a tree of maximum weight given a weighted collection of subsets of some universal set. Here, the weights may indicate lengths of sequences shared among sets of transcripts and the sets that do not share sequences of significant length are assigned zero weights.

Even though the laminar packing problem was motivated by the RNA-seq quantification application described above, it has applications in many other settings. Similar problems have been studied in the context of phylogeny construction where, e.g., Bryant [2] considered the problem of finding a sub-collection of maximum weight given a collection of weighted subsets of a given set, and described a polynomial time algorithm. However, the formulation requires all subsets corresponding to nodes in the tree to be in the collection of subsets whereas the laminar packing problem allows for subsets not in the collection to be included (although the weights will be zero or equivalently we allow the collection of subsets to form a forest).

Day and Sankoff [3] addressed problems of inferring phylogenies by character compatibility where, given a character by object matrix describing a collection of characters, the goal was to find a compatible sub-collection of maximum size. They proved that decision versions of the problems are NP-complete. We note that the Binary Cladistic Compatibility (BCC) problem discussed in their paper reduces to the laminar packing problem.

Here we prove that the laminar packing problem is NP-hard and provide a hardness-to-approximate result. We also give an approximation algorithm as well as an ILP formulation and a greedy algorithm applicable to practical data sets. Finally, we apply our methods to a real data set from the human transcriptome demonstrating solvability of large instances.

2   Preliminaries

In this section, we formally introduce the laminar packing problem. The motivation is to construct a rooted tree or a set of rooted trees (forest) from a set of transcripts where

- The leaves correspond to individual transcripts
- Internal nodes correspond to subsets of transcripts

– Internal nodes have weights reflecting the amount of sequence shared among the subset of transcripts they correspond to

First, for each subset of transcripts, that share sequences with length greater than a threshold, a weight is calculated. This may be done using alignment or using the number of k-mers shared among transcripts. The subsets, with lengths of shared sequence below the threshold, are assigned zero weights implicitly. The goal then is to construct a rooted X-forest with maximum weight.

Definition 1 (Rooted X-Forest). *Let X be a set. A rooted X-forest T is:*

1. *A collection of rooted trees.*
2. *A function $\phi : X \to V(T)$ providing a partial labeling of (non-root) vertices such that every unlabeled vertex has degree > 2.*

A related concept is that of laminar families.

Definition 2 (Laminar Family). *A family, $\mathcal{F}$ of subsets of X is called laminar if and only if for any $S_i, S_j \in \mathcal{F}$, either $S_i \cap S_j = \emptyset$ or $S_i \subset S_j$ or $S_j \subset S_i$.*

Edmonds and Giles established the following connection between laminar families and rooted forests.

Theorem 3 (Edmonds-Giles 1977 [4]). *There is a bijection between laminar families for X and rooted X-forests.*

The laminar packing problem is then defined as follows.

Definition 4 (Laminar Packing Problem). *Given a set X of m items, a collection C of n subsets of X and a weighting function $w : C \to \mathcal{R}^+$, the laminar packing problem is to find a laminar family of maximum weight for X (or equivalently a rooted X-forest of maximum weight).*

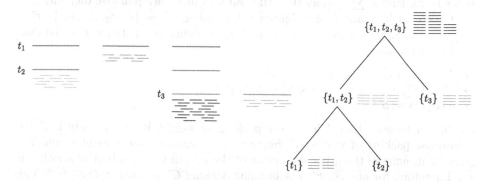

Fig. 1. Three transcripts and reads that map to them are shown on the left. Some reads map uniquely to transcripts while some others map ambiguously. A tree for the three transcripts is shown on the right with uniquely mapped reads assigned to leaves and ambiguously mapped reads to internal nodes

The forest thus constructed may be used for transcript abundance estimation using a streaming EM algorithm where the reads mapping to a set of transcripts can be assigned to the corresponding set instead of a single transcript at the time of processing (Fig. 1). Once all the reads have been processed, the set of trees may be used to get estimates of abundances of transcripts. We note that a forest may not exactly capture the ambiguity structure of transcripts as shown in Fig. 2.

Fig. 2. Figure shows three transcripts whose ambiguity structure is not properly captured by a tree since each of the three pairs has shared segments

3   Complexity of Laminar Packing

The following theorem shows that laminar packing is NP-hard. We reduce the weighted version of the set packing problem to laminar packing. Set packing was proved to be NP-complete by Karp [8].

Theorem 5. *Laminar packing is NP-hard.*

Proof. We will reduce weighted set packing to laminar packing. Given a universe, $U = \{e_1, e_2, \ldots, e_m\}$ and a family $F = \{S_1, S_2, \ldots, S_n\}$ of subsets of U with weights $w_i = w(S_i)$, the weighted set packing problem is to find a subfamily F' so as to maximize $\sum_{S_i \in F'} w_i$ such that all sets in F' are pairwise disjoint.

Given an instance I of weighted set packing, $U = \{e_1, e_2, \ldots, e_m\}$, $F = \{S_1, S_2, \ldots, S_n\}$ with weights $w_i = w(S_i)$, we define an instance I' of laminar packing as follows.

- $X = \{e_1, e_2, \ldots, e_m, t_1, t_2, \ldots, t_n\}$
- $C = \{S'_1, S'_2, \ldots, S'_n\}$ where $S'_i = S_i \cup \{t_i\}$
- $w(S'_i) = w(S_i)$

It can be seen that I has a set packing of weight W if and only if I' has a laminar packing of weight W. Since each S'_i contains an element t_i which is present in none of the other sets, none of the sets in C is a subset of another in C. Therefore, for any S'_i, S'_j in a laminar packing C' we have $S'_i \cap S'_j = \emptyset$. This implies that corresponding sets in F constitute a set packing.

Conversely, we have for any S_i, S_j in a set packing F', $S_i \cap S_j = \emptyset$. This along with the fact that all t_i's are distinct implies that $S'_i \cap S'_j = \emptyset$. Therefore, corresponding S'_i's is a laminar packing. $\square$

An example of a reduction from an instance of set packing to an instance of laminar packing is shown in Fig. 3.

Instance of set packing, I	Instance of laminar packing, I'
$U = \{1, 2, 3, 4, 5, 6\}$	$X = \{1, 2, 3, 4, 5, 6, 7, 8, 9, 10\}$
$F = \{S_1, S_2, S_3, S_4\}$	$C = \{S_1', S_2', S_3', S_4'\}$
$S_1 = \{1, 2, 3\}$	$S_1' = \{1, 2, 3, 7\}$
$S_2 = \{4, 5\}$	$S_2' = \{4, 5, 8\}$
$S_3 = \{1, 2\}$	$S_3' = \{1, 2, 9\}$
$S_4 = \{3, 4\}$	$S_4' = \{3, 4, 10\}$
$w_i = w(S_i)$	$w(S_i') = w_i$

Fig. 3. An example of a reduction from set packing to laminar packing

Set packing is known to be hard to approximate within $n^{(1-\epsilon)}$ and $m^{(1/2-\epsilon)}$ for any $\epsilon > 0$ unless NP = ZPP [6,7]. The above reduction is approximation preserving and the number of sets n remains unchanged in the reduction. Therefore, we have the following corollary.

Corollary 6. *Laminar packing is hard to approximate within $n^{(1-\epsilon)}$ for any $\epsilon > 0$ assuming $NP \neq ZPP$.*

However, the inapproximability result in terms of the number of elements in the universe m does not apply, since the number of elements in the universe increases from m to $m + n$ in the reduction from an instance of set packing to an instance of laminar packing, where n is the number of sets in the family F.

4   Algorithms

The complexity results in Sect. 3 indicate that exact or constant factor approximation algorithms for laminar packing are unlikely. In this section, we show that there is an $O(n/\log^2 n)$ approximation algorithm for the problem and then explore various approaches to solving practical instances of it.

4.1   An Approximation Algorithm

The weighted independent set problem is, given a graph $G = (V, E)$ and weights on vertices, to find a subset $V' \in V$ of maximum weight such that no two vertices in V' are joined by an edge in E. Weighted independent set can be approximated within $O(n/\log^2 n)$ [5].

Theorem 7. *Laminar packing can be approximated within $O(n/\log^2 n)$.*

Proof. An instance of laminar packing I can be transformed into a weighted independent set instance I' as follows. We have a vertex v_i for each set $S_i \in C$

with the weight of v_i equal to the weight of S_i. Two vertices corresponding to sets S_i and S_j are adjacent iff none is a proper subset of the other and $S_i \cap S_j \neq \emptyset$.

We observe that if I has a laminar packing of weight W, I' has a weighted independent set of weight W. If two sets S_i and S_j are in a laminar packing, they are either disjoint or one is a proper subset of the other. This implies that there is no edge between v_i and v_j, and they can be in an independent set. Moreover, the weight of the independent set must equal the weight of the laminar packing.

Conversely, there is no edge between any pair of vertices in an independent set in I'. This implies that corresponding sets in I are either disjoint or one is a proper subset of the other. Therefore, these sets are a laminar family.   □

Figure 4 shows an example of a reduction from an instance of the laminar packing problem to an instance of the independent set problem.

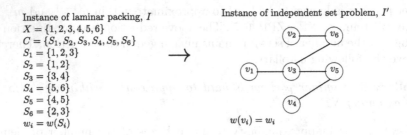

Fig. 4. An example of a reduction from laminar packing to independent set

4.2   An ILP Formulation

The reduction from laminar packing to weighted independent set (WIS), as illustrated in Fig. 4, enables us to solve instances of the laminar packing problem using an integer linear program (ILP) formulation of WIS. There are ILP solvers available that can solve large instances of many NP-hard problems. Given an instance of weighted independent set, $G = (V, E)$ and $w_i = w(v_i)$ for each $v_i \in V$, an ILP formulation is as follows:

$$\text{Maximize} \quad w_1 x_1 + w_2 x_2 + \cdots + w_n x_n$$
$$\text{Such that} \quad x_i + x_j \leq 1 \quad \text{for each } (v_i, v_j) \in E$$
$$x_i \in \{0, 1\} \quad \text{for } i = 1, 2, \ldots, n$$

4.3   A Greedy Algorithm

We also consider a greedy approach to solve instances of the laminar packing problem without converting them to instances of the weighted independent set problem. First, we sort the sets in descending order of weights. We then select sets in sorted order eliminating any set in conflict with the set selected until all sets have been selected or removed. The algorithm is summarized in Algorithm 1. The algorithm runs in $O(n^2 m)$ time, where m is the number of elements in the universe set and n is the number of subsets in the instance.

Algorithm 1. Greedy Laminar Packing (collection of sets C, weights W)

Sort sets in C in descending order of weights
$C' = \emptyset$
while C is not empty **do**
   $S_i \leftarrow$ Set of the highest weight in C
   Add S_i to C' and remove S_i from C
   for each $S_j \in C$ **do**
     if $S_j \not\subseteq S_i$ **and** $S_i \not\subseteq S_j$ **and** $S_i \cap S_j \neq \emptyset$ **then**
       Remove S_j from C
     end if
   end for
end while
return C'

---

5   Experimental Results

We used the human transcriptome (cDNA) downloaded from Ensembl containing 187,626 transcripts as our data set. The sets and their weights are then computed using a combination of C++ and Python codes as follows:

- For each k-mer (contiguous sequence of length k) present in any of the transcripts, we determine the set of transcripts containing the k-mer. We use $k = 31$.
- The weight of a set is then assigned by counting the number of k-mers that appear in that particular set of transcripts.

Table 1 shows some properties of the instance of laminar packing thus constructed and the corresponding instance of the independent set problem. We note that there is one large connected component which makes the strategy of running independent set algorithms on each connected component separately ineffective.

The instance of laminar packing thus obtained was solved using two approaches. The first approach is to convert it to an instance of the weighted independent set problem and then solve an ILP formulation of it using IBM ILOG CPLEX Optimization Studio.

We also apply the greedy algorithm described in Algorithm 1. The greedy algorithm has been implemented in C++. In the implementation, to speed up the search for conflicting sets, we maintain auxiliary lists containing, for each element, the sets that include the element. While searching for conflicts with

Table 1. Properties of laminar packing and corresponding independent set instances

Number of transcripts	187,626
Number of sets (vertices) excluding singletons	546,755
Maximum weight	67,331
Average weight	100.2
Number of edges	101,819,351
Number of connected components	15,471
Size of the largest connected component	444,672
Average size of connected components	35.3
Maximum degree of vertices	59176
Average degree	372.45

some set, it is sufficient to search the lists corresponding to elements in that set, since all other sets are disjoint with the set in question.

A comparison of results obtained through the two approaches is shown in Table 2. Here, solution value refers to the sum of weights of sets of transcripts selected by the ILP and the greedy algorithm. We observe that the greedy algorithm is quite fast and can achieve around 90% of the optimal value.

Table 2. Comparison of results obtained through ILP solver and the greedy algorithm

Approach	Solution value (total weight)	Time
Solving ILP using CPLEX	38,020,248	7 h 39 min 57 s*
Greedy algorithm	34,062,330	4 s

* Excluding time for the conversion to the independent set instance

6   Conclusions

In this paper, we introduced the laminar packing problem, the motivation of which is to stabilize transcript abundance estimates by streaming EM algorithms. We prove that the problem is NP-hard and hard to approximate within $n^{(1-\epsilon)}$ for any $\epsilon > 0$ assuming $NP \neq ZPP$. We also provide an $O(n/\log^2 n)$ approximation algorithm.

We have also explored approaches based on ILP and a greedy strategy to solve large practical instances of the problem and applied our methods to real human transcriptome data set. It can be observed that the greedy algorithm is quite fast and its output is approximately 90% of the optimal value. In the future, other greedy strategies, such as sorting based on weight per degree, may be considered. Finally, it may be incorporated in a tool using the online EM algorithm for transcript abundance estimation to analyze the stability in accuracy provided and the speed-up achieved due to the reduced number of iterations in the online EM algorithm compared to tools employing the offline version.

Acknowledgements. The authors thank Richard Karp, Adam Roberts, and Natth Bejraburnin for helpful conversations throughout the project. ILPs were solved using a free academic distribution of IBM ILOG CPLEX Optimization Studio. LP was funded in part by NIH R01 HG006129. This research was partially conducted at Bangladesh University of Engineering and Technology (BUET) and AR was supported in part by a grant from BUET.

References

1. Bray, N.L., Pimentel, H., Melsted, P., Pachter, L.: Near-optimal probabilistic RNA-Seq quantification. Nat. Biotechnol. **34**(5), 525 (2016)
2. Bryant, D.: Hunting for trees in binary character sets: efficient algorithms for extraction, enumeration, and optimization. J. Comput. Biol. **3**(2), 275–288 (1996)
3. Day, W.H., Sankoff, D.: Computational complexity of inferring phylogenies by compatibility. Syst. Biol. **35**(2), 224–229 (1986)
4. Edmonds, J., Giles, R.: A min-max relation for submodular functions on graphs. In: Hammer, P.L., Johnson, E.L., Korte, B., Nemhauser, G. (eds.) Studies in Integer Programming Annals of Discrete Mathematics, vol. 1, pp. 185–204. Elsevier (1977) https://doi.org/10.1016/S0167-5060(08)70734-9. http://www.sciencedirect.com/science/article/pii/S0167506008707349
5. Halldórsson, M.M.: Approximations of weighted independent set and hereditary subset problems. J. Graph Algorithms Appl. **4**(1), 1–16 (2000)
6. Halldórsson, M.M., Kratochvıl, J., Telle, J.A.: Independent sets with domination constraints. Discrete Appl. Math. **99**(1–3), 39–54 (2000). https://doi.org/10.1016/S0166-218X(99)00124-9. http://www.sciencedirect.com/science/article/pii/S0166218X99001249
7. Håstad, J.: Clique is hard to approximate within $n^{(1-\epsilon)}$. In: Acta Mathematica, pp. 627–636 (1996)
8. Karp, R.M.: Reducibility among combinatorial problems. In: Miller, R.E., Thatcher, J.W., Bohlinger, J.D. (eds.) Complexity of Computer Computations, pp. 85–103. Springer, Heidelberg (1972). https://doi.org/10.1007/978-1-4684-2001-2_9
9. Li, B., Dewey, C.N.: RSEM: accurate transcript quantification from RNA-Seq data with or without a reference genome. BMC Bioinform. **12**(1), 323 (2011)
10. Patro, R., Duggal, G., Love, M.I., Irizarry, R.A., Kingsford, C.: Salmon provides fast and bias-aware quantification of transcript expression. Nat. Methods **14**(4), 417 (2017)
11. Patro, R., Mount, S.M., Kingsford, C.: Sailfish enables alignment-free isoform quantification from RNA-Seq reads using lightweight algorithms. Nat. Biotechnol. **32**(5), 462 (2014)
12. Roberts, A., Pachter, L.: Streaming fragment assignment for real-time analysis of sequencing experiments. Nat. Methods **10**(1), 71 (2013)
13. Trapnell, C., et al.: Transcript assembly and quantification by RNA-Seq reveals unannotated transcripts and isoform switching during cell differentiation. Nat. Biotechnol. **28**(5), 511 (2010)

Efficient Algorithms for Finding Edit-Distance Based Motifs

Peng Xiao, Xingyu Cai, and Sanguthevar Rajasekaran

Department of Computer Science and Engineering,
University of Connecticut, Storrs, CT, USA
{peng.xiao,xingyu.cai,sanguthevar.rajasekaran}@uconn.edu

Abstract. Motif mining is a classical data mining problem which aims to extract relevant information and discover knowledge from voluminous datasets in a variety of domains. Specifically, for the temporal data containing real numbers, it is formulated as time series motif mining (TSMM) problem. If the input is alphabetical and edit-distance is considered, this is called Edit-distance Motif Search (EMS). In EMS, the problem of interest is to find a pattern of length l which occurs with an edit-distance of at most d in each of the input sequences.

There exists some algorithms proposed in the literature to solve EMS problem. However, in terms of challenging instances and large datasets, they are still not efficient. In this paper, EMS3, a motif mining algorithm, that advances the state-of-the-art EMS solvers by exploiting the idea of projection is proposed. Solid theoretical analyses and extensive experiments on commonly used benchmark datasets show that EMS3 is efficient and outperforms the existing state-of-the-art algorithm (EMS2).

Keywords: Sequence analysis · Edit-distance motif · Projection

1 Introduction

Effective data mining algorithms when applied on biological data can reveal crucial information that could lead to accurate diagnosis, drug development, and disease treatment. One set of such mining algorithms are referred to as motif mining (or motif search) algorithm. These algorithms look for information that is closely preserved across species. For example, a piece of gene segment may appear exactly or with minor differences across different species. Extracting such information is very meaningful in numerous applications, such as the determination of open reading frames, identification of gene promoter elements, location of RNA degradation signals, and the identification of alternative splicing sites. Many motif mining models have been proposed.

This paper focuses on the Edit-distance based Motif Search (EMS) model. EMS is defined as follows: Inputs are two integers l and d, and n biological strings over the alphabet Σ of a finite size. Each string is of length m. The problem is to find all the strings of length l that appear in each of the n input strings with

© Springer Nature Switzerland AG 2019
I. Holmes et al. (Eds.): AlCoB 2019, LNBI 11488, pp. 212–223, 2019.
https://doi.org/10.1007/978-3-030-18174-1_16

the Levenshtein distance (or edit-distance) of at most d. Biologists may also be interested in motifs that occur in a fraction of the input strings. The problem of identifying such motifs is known as quorum Edit-distance Motif Search (qEMS). In this case, an extra input parameter q is provided. The problem is to identify all the (l, d, q)-motifs, that is, all (l, d)-motifs that occur in at least qn of the input strings. The standard EMS problem becomes a special case of qEMS when $q = 1$.

In EMS, the edit-distance is used to bound the variability of the pattern across the biological sequences. It can include substitution, insertion and deletion. If only substitution is allowed (*i.e.*, the aim is to find the strings of length l with a Hamming distance of at most d in each of the input sequences), this simplified version of the problem is named as Planted Motif Search (PMS). There are many studies on PMS problems (see *e.g.*, [7,8,16]). EMS is more challenging than PMS because EMS is more general.

It is known that there is a polynomial time reduction from the Closest Substring problem to PMS [8]. Since the Closest Substring problem is NP-Hard [3,4,6], PMS problem is also NP-Hard. EMS is also NP-hard since PMS is a special case of EMS. Therefore, it is of pressing need to develop efficient exact algorithms for EMS problems.

EMS and its variations have been studied since a long time ago. Back in 1998, the authors in [13] proposed an algorithm to find approximate repeats from a long DNA sequence, allowing general insertions and deletions. This is an approximate algorithm. Suffix tree based algorithms are also developed to find approximate repeated or common motifs [14]. The algorithms proposed in [14] can be extended to deal with gaps but the authors did not implement it for edit-distance but only for Hamming distance. The authors in [1] proposed a new algorithm to extract common motifs using the techniques for extracting approximate non-tandem repeats and they also implemented Sagot's algorithm in [14] and did a comparison. The result shows that their algorithm has less false positive motifs and is also more efficient for finding moderately long motifs.

Algorithms proposed in the literature above are part of the early stage studies of the EMS problem. However, the first formal definition of EMS is given in [11] although the authors did not explicitly name it as EMS. They also proposed a deterministic (DMS) algorithm that runs in time $O(n^2 m P^D |\Sigma|^D)$ using $O(nmD + P^D |\Sigma|^D)$ space (P and D are motif length and maximum allowed edit-distance, respectively). A Monte Carlo algorithm with a run time of $O(((n^2 m^2 \log n)/q)D + gmnD)$ is also proposed where g is the number of P-mers that occur in q or more sequences in the database. Following this definition, Pathak *et al.* [10] proposed EMS1 which naturally extends the data structure of d-neighborhood tree from the PMS problem and they evaluated this algorithm on synthetic datasets as well as real datasets. However, one drawback of EMS1 is that it generates too many repeated neighborhoods which takes up a huge memory and also the (l, d) instances it can solve are very limited. To alleviate this problem, Pal *et al.* [9] proposed EMS2. They used wildcard characters to compactly represent the neighborhood tree and also proposed 9 rules to avoid duplications of candidate motifs. EMS2P, which is a parallel algorithm, was also developed and tested on a multi-core machine. Experimental results showed that EMS2 is faster than EMS1 and the parallel version has a good scaling performance.

In this paper, an improved algorithm, EMS3, is proposed to further advance the state-of-the-art EMS solvers. Theoretical study and extensive experimental tests are performed. The rest of the paper is organized as follows. Section 2 presents the proposed algorithm. In Sects. 3 and 4, theoretical analyses and empirical studies of EMS3 are provided. Finally, a brief summary concludes the paper in Sect. 5.

2   EMS3: An Improved Algorithm

2.1   Overview of the Algorithm

EMS3 has 5 steps as follows.

Step 1: Divide Choose an appropriate value of $\epsilon_1 \in (0, 1]$. Let $n' = n * \epsilon_1$. Randomly select n' sequences from the input I. Let this set be I_1. Let $I_2 = I - I_1$.

Step 2: Compress Choose an appropriate value of $\epsilon_2 \in (0, 1]$. Let $|\Sigma'| = |\Sigma| * \epsilon_2$. Compress I_2 by projecting Σ to Σ'. Specifically, every $1/\epsilon_2$ characters in Σ will be projected to a single character in Σ'. For example, if $\Sigma = \{A, C, G, T\}$ and $\epsilon_2 = 1/2$, A, C will be projected to A while G, T will be projected to C. $\Sigma' = \{A, C\}$. This process is also called as encoding.

Step 3: Solve the subproblems Run existing EMS solver on I_1 and I_2. Let the outputs be C_1 and C_2, respectively. Note that the strings from C_2 are in the domain of Σ'. Both C_1 and C_2 are sorted.

Step 4: Merge One idea is to expand the strings in C_2. Let the resultant string set be C_2'. C_2' is in the domain of Σ. The intersection of C_1 and C_2', denoted as C, will be the final candidate set. However, this solution, to a great extent, wipes out the advantage of reducing the alphabet size in the second step because the size of C_2' will be too large. Moreover, C_2' is not sorted any more. Another round of radix sort needs to be performed on C_2' to merge these two sets. A better idea to salvage the execution time is to encode the strings in C_1 in the same way as discussed above. For an l-mer $u \in C_1$, the encoded string is u'. Check if it is in C_2. If it is, add u to C. Note that since C_2 is sorted, a binary search can be performed.

Step 5: Verify Let the output of EMS3 be O. For every l-mer $v \in C$, check if it is an (l, d) edit-distance motif in the remaining sequences, *i.e.*, I_2. If so, add v to O. Three algorithms are proposed for this step.

VerifyMotif_1, *VerifyMotif_2* and *VerifyMotif_3* are the pseudocodes. For all these three algorithms, the inputs are two integers l and d, a set of sequences $\{S_i\}$ $(i = 1, 2, \ldots, n)$, and an l-mer v. The output is a boolean flag indicating whether v is the target (l, d)-motif.

Algorithm 1. VerifyMotif_1(l, d, v, $\{S_i\}$)

$i \leftarrow 1$; $isMotifSeq \leftarrow$ False; $isMotifSeqs \leftarrow$ True;
while $i \leqslant |\{S_i\}|$ **do**

    for $k \leftarrow l - d$ to $l + d$ **do**

        | $subS_i \leftarrow$ the collection of all substrings of length k in S_i;

    for *every* $x \in subS_i$ **do**

        $e(v,x) \leftarrow$ EditDistance(v,x);   // Dynamic programming to compute

        edit-distance between v and x

        if $e(v,x) \leqslant d$ **then**

            | $isMotifSeq \leftarrow$ True;

            | $i \leftarrow i + 1$; Break;

    if $isMotifSeq =$ *False* **then**

        | $isMotifSeqs \leftarrow$ False;

        | Break;

return $isMotifSeqs$;

Algorithm 2. VerifyMotif_2(l, d, v, $\{S_i\}$)

$i \leftarrow 1$; $isMotifSeq \leftarrow$ False; $isMotifSeqs \leftarrow$ True;
$T \leftarrow v$'s d-neighborhood;   // Call GenerateNeighborhoodTree in [9]
while $i \leqslant |\{S_i\}|$ **do**

    for *every* $w \in T$ **do**

        $isMotifSeq \leftarrow$ ExactPatternMatch(w, S_i);   // KMP algorithm [5] to

        check if w appears exactly in S_i

        if $isMotifSeq =$ *True* **then**

            | $i \leftarrow i + 1$; Break;

    if $isMotifSeq =$ *False* **then**

        | $isMotifSeqs \leftarrow$ False;

        | Break;

return $isMotifSeqs$;

Algorithm 3. VerifyMotif_3(l, d, v, $\{S_i\}$)

$i \leftarrow 1$; $isMotifSeq \leftarrow$ False; $isMotifSeqs \leftarrow$ True;
$T \leftarrow v$'s d-neighborhood;   // Call GenerateNeighborhoodTree in [9]
while $i \leqslant |\{S_i\}|$ **do**

    for $k \leftarrow l - d$ to $l + d$ **do**

        | $subS_i \leftarrow$ a collection of substrings of length k in S_i;

    for *every* $x \in subS_i$ **do**

        if $x \in T$ **then**   // Perform binary search

            | $isMotifSeq \leftarrow$ True;

            | $i \leftarrow i + 1$; Break;

    if $isMotifSeq =$ *False* **then**

        | $isMotifSeqs \leftarrow$ False;

        | Break;

return $isMotifSeqs$;

2.2   Why Project the Alphabet

It is desirable to reduce the size of the input while maintaining the accuracy of the algorithm. One way is to project the high dimensional space of the input data into a low dimensional one. The authors in [2] use this technique to find the planted motifs in PMS problems. They randomly choose k selected positions of each l-mer x as a hash function $h(x)$. In other words, they project the motif length from l to k. In [12,16,17], the authors use the idea of random sampling. They randomly select n' out of n sequences and run PMS solvers on the sample dataset. This can be considered as a projection of the number of biological sequences from n to n'.

To the best of the authors' knowledge, the idea of alphabet projection has not been employed before to solve motif mining problems. It is believed that the alphabet size has a great impact on the time complexity of EMS algorithms. For example, in [11], the authors proposed an algorithm to solve the qEMS problem that runs in $O(n^2 m l^d |\Sigma|^d)$. In [10], the authors proposed EMS1 which has a time complexity of $O(mn(4l|\Sigma|)^d + |\Sigma|^l)$. Compared to sampling n, projecting the alphabet to a smaller size will greatly reduce the running time.

2.3   Correctness of the Algorithm

It is easy to see that EMS3 is a deterministic algorithm that always output the correct motifs. An important question is whether EMS3 misses any true motifs. The answer is no. Assume that the set of true motifs is G. It can be proved that after the merge step, the candidate motif set C is a superset of the true motif set G, $i.e.$, $G \subseteq C$.

Please note due to page limit, from this point, proof of the lemmas and theorems are omitted. Interested readers can ask the authors for details.

Lemma 1. *Let l_1 and l_2 be strings on Σ and let the edit-distance between l_1 and l_2 be d. Let l'_1 and l'_2 be compressed strings of l_1 and l_2 using the projection technique discussed above from Σ to Σ' ($|\Sigma'| \leqslant |\Sigma|$). The edit-distance between l'_1 and l'_2, denoted as d', is no more than d.*

Theorem 2. $G \subseteq C$.

3   Analysis of EMS3

3.1   Time Complexity Analysis

Expected Number of Candidate Motifs. The expected number E of candidate motifs is a function of $m, n, l, d, |\Sigma|$, and is derived in [9], to which the interested reader is referred for details. In that paper, the Eqs. 1, 2 and 3 below, with δ, β, α, and q acting as dummy variables, lead to an expression for E in Eq. 4.

$$N(\delta, \beta, \alpha, l, |\Sigma|, q) = \binom{l+q}{\delta}\binom{l+q-\delta}{\beta}\binom{l+q-\delta+\alpha}{\alpha}|\Sigma|^\alpha(|\Sigma|-1)^\beta. \quad (1)$$

$$P(l, |\Sigma|, d, q) = \sum_{\delta=\max\{0,q\}}^{\frac{d+q}{2}} \frac{N(\delta, d+q-2\delta, \delta-q, l, |\Sigma|, q)}{|\Sigma|^{l+q}}. \tag{2}$$

$$R(m, l, d, |\Sigma|) = \Pi_{q=-d}^{d}(1-P)^{m-l-q+1}. \tag{3}$$

$$E(m, n, l, d, |\Sigma|) = |\Sigma|^{l}(1-R)^{n}. \tag{4}$$

The expected sizes of C_1 and C_2 can be written as:

$$E(|C_1|) = E(m, n', l, d, |\Sigma|). \tag{5}$$

$$E(|C_2|) = E(m, (n-n'), l, d, |\Sigma'|). \tag{6}$$

An l-mer has a probability of $p_1 = (1 - R(m, l, d, |\Sigma|))^{n'}$ to be in C_1. If it is encoded, it has a probability of $p_2 = (1 - R(m, l, d, |\Sigma'|))^{n-n'}$ to be in C_2. Every $(1/\epsilon_2)^l$ l-mers in Σ will be projected to a single l-mer in Σ'. Therefore, the expected number of l-mers in C is:

$$E(|C|) = |\Sigma|^{l} p_1 p_2 / \epsilon_2^l. \tag{7}$$

In order to make sure that the expected number of candidate motifs is reduced, it is desirable to have:

$$p_2/\epsilon_2^l < 1. \tag{8}$$

However, it does not mean that p_2/ϵ_2^l should be as small as possible. When ϵ_2 is large, the size of the candidate motif set is small thus reducing the running time in Step 5 of EMS3. However, the running time of Step 3 increases because of a relatively large alphabet size.

Time Complexity of EMS2. Time complexity of EMS2 is not given in its original paper [9]. It can be shown that the overall time complexity of EMS2 is:

$$T_{EMS2} = O(mndl^{d+1}|\Sigma|^d). \tag{9}$$

Note that this may be larger than the time complexity in [10,11]. This is because unlike [10,11], in EMS2 and EMS3, assumption that the motifs of interest should come exactly from one of the input sequences is removed to make this problem more general. However, this only represents the worst time complexity. A lot of branches of the neighborhood tree can be pruned because of the rules proposed in [9]. Therefore, the actual running time is much less. But it is hard to estimate how many branches will be pruned.

Time Complexity of Verifying Candidate Motifs. There are three algorithms to verify the candidate motifs. The first algorithm uses dynamic programming to compute the edit-distance between an l'-mer $(l - d \leqslant l' \leqslant l + d)$ and an l-mer. Therefore the time taken is:

$$T_{verify\_1} = |C|n \sum_{l'=l-d}^{l+d} O(l'l(m - l' + 1)) = O(|C|mndl^2).$$

$|C|$ is the size of the candidate motif set. An expected number of the candidate motifs $E(|C|)$ can be found in Eq. 7.

The second algorithm will generate the d-neighborhood tree and use a linear time complexity algorithm to locate the neighborhoods within the input sequence. It is known that the number of d-neighborhoods of an l-mer is $O(l^d|\Sigma|^d)$. Thus, time taken in this algorithm is (assuming $d < l < m$):

$$T_{verify\_2} = |C|nO(l^d|\Sigma|^d(m + l)) = O(|C|mnl^d|\Sigma|^d).$$

The third algorithm also generates d-neighborhood tree but tries to locate the k-mer $(l - d \leqslant k \leqslant l + d)$ from the input sequences in the tree. Time complexity for this algorithm is:

$$T_{verify\_3} = |C|nO(\sum_{k=l-d}^{l+d} (m - k + 1)\log(l^d|\Sigma|^d)) = O(|C|mnd^2(\log l + \log |\Sigma|)).$$

These are only upper bounds of the time taken in each algorithm. It looks like that the second algorithm takes the longest time. However, in generating the neighborhood tree, a lot of branches are pruned. The neighborhoods are also sorted and duplications are removed. Therefore the actual number of neighborhoods is far less than $l^d|\Sigma|^d$.

Time Complexity of EMS3. Step 1 and 2 take time that is negligible. Time complexities of Step 3 and 5 are already analyzed. Step 4 takes time $O(l|C_1|\log |C_2|)$. An expected number of $E(|C_1|)$ and $E(|C_2|)$ can be found in Eqs. 5 and 6. Therefore, assuming using *VerifyMotif_1* in the final step to verify the candidate motifs, the time complexity of EMS3 is:

$$T_{EMS3} = O(mdl^{d+1}(n'|\Sigma|^d + (n - n')|\Sigma'|^d) + l|C_1|\log |C_2| + |C|mndl^2). \quad (10)$$

where $n' = n * \epsilon_1, |\Sigma'| = |\Sigma| * \epsilon_2$.

3.2   How to Choose ϵ_1 and ϵ_2

Equation 8 shows the first guideline to choose ϵ_1 and ϵ_2. Generally, in a divide and conquer algorithm, it is desirable to split the input into nearly equal halves so that the performance of the algorithm is the best. However, in EMS3, in order to balance the run times of the two subproblems, their running times should be in the same scale. Therefore, ϵ_1 is smaller than $1/2$. Solve $n'|\Sigma|^d = \Theta((n-n')|\Sigma'|^d)$:

$$\epsilon_1 = \Theta((1 - \epsilon_1)\epsilon_2^d). \tag{11}$$

It is also noteworthy to point out that in conventional notion of divide and conquer, the same algorithm is applied recursively to the subproblems whereas in EMS3, the subproblems are solved non-recursively using EMS2 solver.

3.3   A Discussion on qEMS and Approximate EMS3

EMS3 can be extended to qEMS problems as well. Under such circumstances, besides ϵ_1 and ϵ_2, another parameter ϵ_3 $(0 < \epsilon_3 \leqslant 1)$ is needed.

Theorem 3. *If an l-mer x is an (l, d, q) edit-distance motif on Σ, then the encoded l-mer x' is also an (l, d, q) edit-distance motif on Σ' using the projection technique as discussed, with $|\Sigma'| \leqslant |\Sigma|$.*

Theorem 4. *If an l-mer x is an (l, d, q) edit-distance motif on n sequences, then it is also an $(l, d, q\epsilon_3)$ edit-distance motif on $n\epsilon_1$ $(0 < \epsilon_1 \leqslant 1)$ sequences as long as the following condition is satisfied: $\epsilon_1(1 - \epsilon_3 q) \geqslant 1 - q$.*

Theorem 5. *If an l-mer x is an (l, d, q) edit-distance motif on n sequences, then it is also an $(l, d, q\epsilon_3)$ edit-distance motif on $n\epsilon_1$ $(0 < \epsilon_1 \leqslant 1)$ sequences with a high probability as long as: $(1 - \epsilon_3)^2\epsilon_1 \geqslant \frac{2\beta \ln(n)}{nq}$. β is a constant. A high probability means that the probability is higher than $1 - n^{-\beta}$.*

When n is small, ϵ_1 and ϵ_3 can be chosen according to Theorem 4 to make sure the algorithm can always output the correct answer. When n is large, ϵ_1 and ϵ_3 can be chosen according to Theorem 5. In this case, EMS3 becomes a randomized algorithm.

If Step 5 of EMS3 is removed, EMS3 becomes an approximate algorithm. It is known that the candidate motifs set C is a superset of the true motifs set G.

Theorem 6. *If Step 5 of EMS3 is removed, then it becomes an approximate algorithm with an expected approximation ratio of $E(|C|)/E(m, n, l, d, |\Sigma|)$. $E(|C|)$ and $E(m, n, l, d, |\Sigma|)$ can be found in Eqs. 7 and 4, respectively.*

4   Experimental Evaluations of EMS3

Extensive experiments on existing standard benchmark datasets are performed to evaluate EMS3. All the algorithms are evaluated on a Dell Precisions Workstation T7910 running RHEL 7.0 on two sockets each containing 8 Dual Intel Xeon Processors E5-2667 (8C HT, 20 MB Cache, 3.2 GHz) and 256 GB RAM. *VerifyMotif_1* is used as the algorithm to verify the candidate motifs.

Table 1. Size of candidate motif sets with different n' and $|\Sigma|'$

(a) $(l,d) = (8,1)$			(b) $(l,d) = (12,2)$			(c) $(l,d) = (16,3)$		
n'	$|C_1|$	$\dfrac{|C|}{|\Sigma|'=2 \;\; |\Sigma|'=3}$	n'	$|C_1|$	$\dfrac{|C|}{|\Sigma|'=2 \;\; |\Sigma|'=3}$	n'	$|C_1|$	$\dfrac{|C|}{|\Sigma|'=2 \;\; |\Sigma|'=3}$
2	6762	6762 4549	2	72763	72749 21157	2	721497	719577 18794
4	667	667 503	4	389	389 99	4	203	203 7
6	74	74 58	6	9	9 5	6	1	1 1
8	7	7 5	8	1	1 1	8	1	1 1
10	2	2 1	10	1	1 1	10	1	1 1

4.1 Synthetic Datasets

Following the tradition, n ($= 20$) DNA sequences of length m ($= 600$) each are generated, where each character is independent and identically distributed (i.i.d.) over the alphabet under concern ($\Sigma = \{A, C, G, T\}$). A random string M of length l is randomly generated as the target motif. Besides, a d-neighborhood string is planted in each of the n sequences. In addition to the motif planted, there could be other motifs that occur by random chance. Challenging instances of $(l,d) = (8,1), (12,2), (16,3)$ are tested. $(l,d) = (20,4)$ is not tested because EMS2 cannot complete it within stipulated 72 h.

n' varies from 2 to 10. $|\Sigma|' = 2$ and 3. From Table 1, it can be seen that when $|\Sigma|' = 2$, the size of candidate motif set $|C|$ is not reduced. In fact, it is almost exactly the same as $|C_1|$. Therefore, as discussed before, it does not mean a large compression ratio is necessarily better. However, when $|\Sigma|' = 3$, $|C|$ is much smaller than $|C_1|$. It concurs with the analysis in Sect. 3. It is wise to pick a relatively small value of ϵ_1 and a relatively large value of ϵ_2. For example, when $(l,d) = (16,3)$, setting $|\Sigma|' = 3$ and $n' = 4$ will reduce the size of candidate set from 203 to 7.

The running time of EMS2 for $(l,d) = (8,1), (12,2), (16,3)$ are 0.14 s, 14.86 s and 21.18 m, respectively. Table 2 shows that generally EMS3 has a good speedup over EMS2. The best speedups for $(l,d) = (8,1), (12,2), (16,3)$ are 1.75, 1.68 and 1.84 when $n' = 8$ (or 10), 4, 2. Figure 1 shows the speedups of EMS3 over EMS2 with different n' on the challenging instances. When n' is chosen appropriately, EMS3 is expected to have around 70% or 80% improvement in speed. When the alphabet size or the number of sequences is larger, EMS3 is expected to perform much better than EMS2. There will also be more choices for picking the parameters.

4.2 Real Biological Datasets

EMS3 is also compared against EMS2 on the real biological DNA datasets ($\Sigma = \{A, C, G, T\}$) discussed in [7,15]. The datasets can be downloaded from http://bio.cs.washington.edu/assessment/download.html.

Table 2. Running time of EMS3 with different n' ($|\Sigma|' = 3$)

n'	(l, d)		
	$(8, 1)$	$(12, 2)$	$(16, 3)$
2	5.83 s	1.24 m	**11.54 m**
4	0.72 s	**8.83 s**	12.70 m
6	0.15 s	10.15 s	14.61 m
8	**0.08 s**	11.67 s	17.22 m
10	**0.08 s**	13.45 s	19.87 m

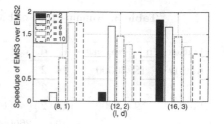

Fig. 1. Speedups of EMS3 over EMS2 with different (l, d) and n'

The "real" benchmark datasets (file names with suffix r) which have the binding sites in their real genomic promoter sequences are chosen as the test files. Datasets with less than 8 input sequences are excluded because they are not very challenging. For each dataset, d is set to 2 and 3. l is chosen on a dataset basis to ensure that the number of reported motifs is not excessive but the instance is challenging as well. When running the experiment of EMS3, instead of exhaustedly varying the parameters of ϵ_1 and ϵ_2, only one combination of ϵ_1 and ϵ_2 is tested. However, in the real datasets, no assumption should be made on the statistical distribution. Therefore, the second guideline can be utilized here to choose the parameters. For example, manually set $\epsilon_2 = 3/4$ and $\epsilon_1 = (1-\epsilon_1)\epsilon_2^d$ and solve it for $\epsilon_1 = 9/25$ when $d = 2$ and $\epsilon_1 = 27/91$ when $d = 3$.

In Table 3, the dataset name, the total number of sequences, the total number of bases in each dataset, the l and d combination, size of candidate motif set when running EMS3, the runtimes of the two algorithms and the speedup of EMS3 over EMS2 are reported. From this table, it is obvious that size of the candidate motif set is generally greatly reduced (*i.e.*, $|C| < |C_1|$). This shows the necessity and effectiveness of pruning $|C_1|$ by checking if the pattern in $|C_1|$, after encoding, can survive in $|C_2|$. When $d = 2$, the improvement in speed is around 30% to 60%, but in rare cases, it only performs slightly better than EMS2. It may be because for some small instances, the overhead brought by EMS3 more or less balances out its advantage. However, when $d = 3$, the improvement in speed is generally over 50% with the maximum speedup of 2.1 when $(l, d) = (17, 3)$ on mus11r dataset.

4.3   Summary of Experimental Evaluation

EMS3 outperforms EMS2 on both synthetic and real datasets. EMS3 works well with challenging instances. This is because the size of the candidate motif set is relatively small thus it will not take much time to verify the candidate motifs. If there is only one motif found which is the planted motif, then EMS3 is very good at capturing this one. EMS3 works better for large datasets and instances. It is expected EMS3 will perform better on protein data and datasets with more input sequences. This is proved in the time complexity analysis above. However, the corresponding experiments are not carried out because EMS2 consumes more than 500 GB memory for protein data even when $d = 3$.

Table 3. Running time of EMS3 over EMS2 on real datasets

Dataset	n	No. bases	l	d	$|C_1|$	$|C|$	T_{EMS3} (s)	T_{EMS2} (s)	Speedup
hm01r	18	36000	14	2	1	0	65.63	89.93	1.37
			18	3	0	0	5774.66	9104.86	1.58
hm02r	9	9000	15	2	68	7	19.44	23.58	1.21
			19	3	266059	1974	1260.21	2065.53	1.64
hm03r	10	15000	15	2	219	18	32.41	42.56	1.31
			19	3	196662	2079	2288.51	3931.03	1.72
hm04r	13	26000	14	2	349	102	44.45	61.27	1.38
			18	3	14938	2338	3232.39	6280.58	1.94
hm08r	15	7500	13	2	5	0	8.07	12.16	1.51
			17	3	1	0	570.29	1107.60	1.94
hm20r	35	70000	13	2	10	7	82.63	132.36	1.60
			17	3	0	0	8809.21	12406.11	1.41
hm26r	9	9000	15	2	220	51	17.32	23.60	1.36
			19	3	105908	579	1121.93	2152.78	1.92
mus02r	9	9000	15	2	178	75	17.40	23.85	1.37
			19	3	230537	3285	1455.39	2026.11	1.39
mus11r	12	6000	13	2	639	15	6.14	9.75	1.59
			17	3	3443	27	417.52	878.54	2.10
yst01r	9	9000	15	2	154	9	17.03	23.62	1.39
			19	3	213654	3920	1278.39	1969.11	1.54
yst03r	8	4000	14	2	8462	421	6.92	7.28	1.05
			19	3	30398	5	441.78	830.92	1.88
yst08r	11	11000	14	2	3603	510	20.91	23.96	1.15
			18	3	20359	831	1432.57	2093.48	1.46
yst09r	169	16000	13	2	96	66	17.17	27.89	1.62
			17	3	784	215	1560.99	2483.25	1.59

5   Conclusions and Future Work

In this paper, EMS3, an improved algorithm is proposed to efficiently solve the EMS problem. EMS3 is a non-recursive divide and conquer algorithm and uses the idea of projection. Theoretical analysis shows that EMS3 is even more competitive for large datasets and challenging instances. The experimental results reveal that EMS3 outperforms EMS2 which is the state-of-the-art algorithm.

In future the authors plan to improve the performance of EMS solvers by reducing the memory usage and focusing on larger datasets. Quorum support can be added to the existing EMS solver. How to project l, d or m in EMS problems is worth considering as well. Besides, developing efficient approximate and randomized algorithms for the EMS problem is also interesting.

Acknowledgment. This work has been supported in part by the NSF grants 1447711, 1743418, and 1843025.

References

1. Adebiyi, E.F., Kaufmann, M.: Extracting common motifs under the levenshtein measure: theory and experimentation. In: Guigó, R., Gusfield, D. (eds.) WABI 2002. LNCS, vol. 2452, pp. 140–156. Springer, Heidelberg (2002). https://doi.org/10.1007/3-540-45784-4_11
2. Buhler, J., Tompa, M.: Finding motifs using random projections. In: Proceedings of Fifth Annual International Conference on Computational Molecular Biology (RECOMB) (2001)
3. Cai, X., Mamun, A.A., Rajasekaran, S.: Novel algorithms for finding the closest l-mers in biological data. In: 2017 IEEE International Conference on Bioinformatics and Biomedicine (BIBM), pp. 525–528. IEEE (2017)
4. Cai, X., Zhou, S., Rajasekaran, S.: Jump: a fast deterministic algorithm to find the closest pair of subsequences. In: Proceedings of the 2018 SIAM International Conference on Data Mining, pp. 73–80. SIAM (2018)
5. Knuth, D.E., Morris Jr., J.H., Pratt, V.R.: Fast pattern matching in strings. SIAM J. Comput. **6**(2), 323–350 (1977)
6. Lanctot, J.K., Li, M., Ma, B., Wang, S., Zhang, L.: Distinguishing string selection problems. Inf. Comput. **185**(1), 41–55 (2003)
7. Nicolae, M., Rajasekaran, S.: Efficient sequential and parallel algorithms for planted motif search. BMC Bioinform. **15**(1), 1 (2014)
8. Nicolae, M., Rajasekaran, S.: qPMS9: an efficient algorithm for quorum planted motif search. Sci. Rep. **5**, 7813 (2015)
9. Pal, S., Xiao, P., Rajasekaran, S.: Efficient sequential and parallel algorithms for finding edit distance based motifs. BMC Genomics **17**(4), 465 (2016)
10. Pathak, S., Rajasekaran, S., Nicolae, M.: EMS1: an elegant algorithm for edit distance based motif search. Int. J. Found. Comput. Sci. **24**(04), 473–486 (2013)
11. Rajasekaran, S., et al.: High-performance exact algorithms for motif search. J. Clin. Monit. Comput. **19**(4–5), 319–328 (2005)
12. Rajasekaran, S., Dinh, H.: A speedup technique for (l, d)-motif finding algorithms. BMC Res. Notes **4**(1), 54 (2011)
13. Rocke, E., Tompa, M.: An algorithm for finding novel gapped motifs in DNA sequences. In: Proceedings of the Second Annual International Conference on Computational Molecular Biology, pp. 228–233. ACM (1998)
14. Sagot, M.-F.: Spelling approximate repeated or common motifs using a suffix tree. In: Lucchesi, C.L., Moura, A.V. (eds.) LATIN 1998. LNCS, vol. 1380, pp. 374–390. Springer, Heidelberg (1998). https://doi.org/10.1007/BFb0054337
15. Tompa, M., et al.: Assessing computational tools for the discovery of transcription factor binding sites. Nat. Biotechnol. **23**(1), 137 (2005)
16. Xiao, P., Pal, S., Rajasekaran, S.: qPMS10: a randomized algorithm for efficiently solving quorum Planted Motif Search problem. In: 2016 IEEE International Conference on Bioinformatics and Biomedicine (BIBM), pp. 670–675. IEEE (2016)
17. Xiao, P., Pal, S., Rajasekaran, S.: Randomised sequential and parallel algorithms for efficient quorum planted motif search. Int. J. Data Min. Bioinform. **18**(2), 124 (2017)

Author Index

Alekseyev, Max A.   97
Atamanova, Maria   97
Avdeyev, Pavel   97

Cai, Xingyu   212
Chateau, Annie   25
Chindelevitch, Leonid   137, 152

Davot, Tom   25

Giroudeau, Rodolphe   25
Guerrini, Veronica   112

Hayes, Wayne   52

Intosalmi, Jukka   191

Kahramanoğulları, Ozan   39
Korobeynikov, Anton   80
Krishnamoorty, Sriram   125
Kristensen, Vessela N.   179
Kulkarni, Milind   125

Lähdesmäki, Harri   191
Le, Thien   167
Levy-Jurgenson, Alona   179

Maharaj, Sridevi   52
Meidanis, João   137, 152
Molloy, Erin K.   167

Nousiainen, Kari   191

Ohiba, Zarin   52

Pachter, Lior   203
Petingi, Louis   68

Rahman, Atif   203
Rajasekaran, Sanguthevar   212
Rao, Satish   167
Rosone, Giovanna   112

Schlick, Tamar   68
Shlemov, Alexander   80
Sy, Aaron   167

Tekpli, Xavier   179

Warnow, Tandy   3, 167
Weller, Mathias   25
Wright, Christopher   125

Xiao, Peng   212

Yakhini, Zohar   179

Zanetti, João Paulo Pereira   137, 152
Zhang, Qiuyi (Richard)   167